Hybrid Metal Additive Manufacturing

The text presents the latest research and development, technical challenges, and future directions in the field of hybrid metal additive manufacturing. It further discusses the modeling of hybrid additive manufacturing processes for metals, hybrid additive manufacturing of composite materials, and low-carbon hybrid additive manufacturing processes.

THIS BOOK

- Presents cutting-edge advancements and limitations in hybrid additive manufacturing technologies.
- Discusses fabrication methods and rapid tooling techniques focusing on metals, composites, and alloys.
- Highlights the importance of low-carbon additive manufacturing technologies toward achieving sustainability.
- Emphasizes the challenges and solutions for integrating additive manufacturing and Industry 4.0 to enable rapid manufacturing of customized and tailored products.
- Covers hybrid additive manufacturing of composite materials and additive manufacturing for fabricating high-hardness components.

The text discusses the recent advancements in additive manufacturing of high-hardness components and covers important engineering materials such as metals, alloys, and composites. It further highlights defects and post-processing of hybrid additive manufacturing components, sustainability solutions for hybrid additive manufacturing processes, and recycling of machining waste into metal powder feedstock. It will serve as an ideal reference text for senior undergraduate and graduate students, and researchers in fields including mechanical engineering, aerospace engineering, manufacturing engineering, and production engineering.

Advances in Manufacturing, Design and Computational Intelligence Techniques

Series Editor:
Ashwani Kumar- Senior Lecturer, Mechanical Engineering, at Technical Education Department, Uttar Pradesh, Kanpur, India

The book series editor is inviting edited, reference and text book proposal submission in the book series. The main objective of this book series is to provide researchers a platform to present state of the art innovations, research related to advanced materials applications, cutting edge manufacturing techniques, innovative design and computational intelligence methods used for solving nonlinear problems of engineering. The series includes a comprehensive range of topics and its application in engineering areas such as additive manufacturing, nanomanufacturing, micromachining, biodegradable composites, material synthesis and processing, energy materials, polymers and soft matter, nonlinear dynamics, dynamics of complex systems, MEMS, green and sustainable technologies, vibration control, AI in power station, analog-digital hybrid modulation, advancement in inverter technology, adaptive piezoelectric energy harvesting circuit, contactless energy transfer system, energy efficient motors, bioinformatics, computer aided inspection planning, hybrid electrical vehicle, autonomous vehicle, object identification, machine intelligence, deep learning, control-robotics-automation, knowledge based simulation, biomedical imaging, image processing and visualization. This book series compiled all aspects of manufacturing, design and computational intelligence techniques from fundamental principles to current advanced concepts.

Thermal Energy Systems: Design, Computational Techniques, and Applications
Edited by Ashwani Kumar, Varun Pratap Singh, Chandan Swaroop Meena, Nitesh Dutt

Hybrid Metal Additive Manufacturing: Technology and Applications
Edited by Parnika Shrivastava, Anil Dhanola, and Kishor Kumar Gajrani

https://www.routledge.com/Advances-in-Manufacturing-Design-and-Computational-Intelligence-Techniques/book-series/CRCAIMDCIT?publishedFilter=alltitles&pd=published,forthcoming&pg=1&pp=12&so=pub&view=list?publishedFilter=alltitles&pd=published,forthcoming&pg=1&pp=12&so=pub&view=list

Hybrid Metal Additive Manufacturing
Technology and Applications

Edited by
Parnika Shrivastava, Anil Dhanola,
and Kishor Kumar Gajrani

CRC Press
Taylor & Francis Group
Boca Raton London New York

CRC Press is an imprint of the
Taylor & Francis Group, an **informa** business

Designed cover image: © Shutterstock

First edition published 2024
by CRC Press
2385 Executive Center Drive, Suite 320, Boca Raton FL 33431

and by CRC Press
4 Park Square, Milton Park, Abingdon, Oxon, OX14 4RN

CRC Press is an imprint of Taylor & Francis Group, LLC

© 2024 selection and editorial matter, Parnika Shrivastava, Anil Dhanola, and Kishor Kumar Gajrani individual chapters,

ISBN: 9781032460550 (hbk)
ISBN: 9781032523996 (pbk)
ISBN: 9781003406488 (ebk)

DOI: 10.1201/9781003406488

Typeset in Sabon
by codeMantra

Contents

P M ABHILASH, JIBIN BOBAN, AFZAAL AHMED, AND XICHUN LUO

13 Additive manufacturing for society   222

ALEX Y, NIDHIN DIVAKARAN, AND SMITA MOHANTY

Acknowledgments

We would like to thank all the authors of various book chapters for their contributions. We extend our heartfelt gratitude to *CRC Press (Taylor & Francis Group)* and the editorial team for their support during the completion of this book. We are sincerely grateful to reviewers for their suggestions and illuminating views on each book chapter presented in the book *Hybrid Metal Additive Manufacturing: Technology and Applications.*

About the editors

Dr. Parnika Shrivastava is an Assistant Professor in the Department of Mechanical Engineering at the National Institute of Technology, Jalandhar, Punjab, India. She earned her Doctorate and M.Tech. degrees in Mechanical Engineering from PDPM, Indian Institute of Information Technology, Design & Manufacturing Jabalpur, Jabalpur, in 2019. She received her B.Tech. degree (Honors) from Mechanical Engineering Discipline of Engineering College Bikaner (E.C.B), Bikaner, Rajasthan. Her research areas of interest are advanced forming operations, hybrid manufacturing, fracture mechanics, rapid product development technologies, and topology optimization. She has been granted a patent on the "Process of analyzing the effect of preheated microstructure vis-à-vis parameters on the orange peel in incremental sheet forming of AA1050 sheets." She has authored various research articles in international journals of repute, along with several conference publications and book chapters. She has been recently granted a Core Research Grant funding for undertaking the project "Design and Development of an Electrically Assisted Hybrid Double Sided Incremental Forming Machine for Efficient Production of Customized Biomedical Implants with Improved Surface Finish and Geometrical Accuracy." She is also the recipient of two national-level awards, including the Best Teacher Award 2022 conferred by the Indian Society for Technical Education, New Delhi.

Dr. Anil Dhanola is an Assistant Professor in the Department of Mechanical Engineering at Chandigarh University, Mohali. He completed his M.Tech. with honors in Production Engineering at Govind Ballabh Pant Institute of Engineering and Technology, Pauri Garhwal in 2015. Dr. Anil earned a full-time Ph.D. in Mechanical Engineering from the Guru Jambheshwar University of Science and Technology, Hisar, Haryana in 2021. He has more than five years of experience in teaching and industry. His domains of research include tribology, fiber-based polymer composites, fluid film lubrication, green lubricants, wind turbine tribology, and superlubricity. Dr. Anil has published several articles in international peer-reviewed journals and book chapters. Dr. Anil has also presented various articles at national and international conferences. He is also a reviewer of various peer-reviewed international journals.

Dr. Kishor Kumar Gajrani is an Assistant Professor in the Department of Mechanical Engineering at the Indian Institute of Information Technology, Design and Manufacturing, Kancheepuram, Chennai, India. He earned his M.Tech. and Ph.D. from the Department of Mechanical Engineering at the Indian Institute of Technology, Guwahati. Thereafter, he worked as a post-doctoral researcher at the Indian Institute of Technology, Bombay. He has authored/co-authored 45+ international journals and book chapters of repute. He has also edited/co-edited three books: *Advances in Sustainable Machining, Biodegradable Composited for Packaging Applications*, and *Sustainable Materials and Manufacturing Technologies*. He was listed in the *World's Top 2% Scientists* list in 2022 released by Stanford University and published by Elsevier. Dr. Gajrani works on the advancement of sustainable machining processes, additive manufacturing, advanced materials, tribology, coatings, green lubricants, and coolants as well as food packaging.

Contributors

Afzaal Ahmed
Department of Mechanical
  Engineering
Indian Institute of Technology
Palakkad, India

Arun Kumar Bambam
Department of Mechanical
  Engineering
Indian Institute of Information
  Technology, Design and
  Manufacturing
Chennai, India

Prabal Batra
Chandigarh University Gharuan
Mohali, India

Shatarupa Biswas
Department of Mechanical
  Engineering
NIT
Silchar, India

Jibin Boban
Mikrotools Pte Ltd.
Jalan Bukit Merah,
  Singapore

Tushar R. Dandekar
School of Mechanical and Design
  Engineering
University of Portsmouth
Hampshire, United Kingdom

Santanu Das
Department of Mechanical
  Engineering
KGEC
Nadia, India

Anil Dhanola
Department of Mechanical
  Engineering
Chandigarh University Gharuan
Mohali, India

Nidhin Divakaran
Laboratory for Advanced Research
  in Polymeric Materials
  (LARPM)
School for Advanced Research in
  Polymers (SARP)
Central Institute of Petrochemicals
  Engineering & Technology
  (CIPET)
Bhubaneswar, India

Kishor Kumar Gajrani
Department of Mechanical
    Engineering
Indian Institute of Information
    Technology, Design and
    Manufacturing
Chennai, India

A Gopichand
Department of Mechanical
    Engineering
Swarnandhra College of
    Engineering and Technology
Narsapur, India

Tuhina Goswami
Department of Mechanical
    Engineering
KGEC
Nadia, India

Jagdeep Kaur
Chandigarh University
Mohali, India

Tarveen Kaur
Chandigarh University
Mohali, India

Avinash Kumar
Department of Mechanical
    Engineering
Indian Institute of Information
    Technology, Design and
    Manufacturing
Chennai, India

Praveen Kumar
Department of Mechanical
    Engineering
Sant Longowal Institute of
    Engineering and Technology
Sangrur, India

Xichun Luo
Centre for Precision Manufacturing
DMEM, University of Strathclyde
Glasgow, United Kingdom

Francis Luther King M
Department of Mechanical
    Engineering
Swarnandhra College of
    Engineering and Technology
Narsapur, India

Smita Mohanty
Laboratory for Advanced Research
    in Polymeric Materials
    (LARPM), School for Advanced
    Research in Polymers (SARP)
Central Institute of Petrochemicals
    Engineering & Technology
    (CIPET)
Bhubaneswar, India

Sachin Moond
Chandigarh University
Mohali, India

Manidipto Mukherjee
CSIR-CMERI
Durgapur, India

Rajesh Jesudoss Hynes N
Faculty of Mechanical Engineering
Opole University of Technology
Opole, Poland
and
Department of Mechanical
    Engineering
Mepco Schlenk Engineering
    College
Sivakasi, India

Shenbaga Velu P
School of Mechanical Engineering
Vellore Institute of Technology,
    Chennai Campus
Chennai, India

Shenbaga Velu P
School of Mechanical Engineering
Vellore Institute of Technology
Chennai, India

Abhilash P. M.
Centre for Precision
    Manufacturing, DMEM
University of Strathclyde
Glasgow, United Kingdom

C. Pradeepkumar
Department of Automobile
    Engineering
Kalasalingam Academy of Research
    and Education
Srivilliputhur, India

Tharmaraj R
Department of Mechanical
    Engineering
SRM Institute of Science and
    Technology, Ramapuram
    Campus
Chennai, India

Tharmaraj R
Department of Mechanical
    Engineering
SRM Institute of Science and
    Technology, Ramapuram
    Campus
Chennai, India

Atri Rathore
Chandigarh University
Mohali, India

Carlo Santulli
School of Science and Technology
Università degli Studi di Camerino
Camerino, Italy

S. Shiva
Laboratory for Advanced
    Manufacturing and Processing
    (LAMP)
Indian Institute of Technology
Jammu, India

G Robert Singh
Department of Mechanical
    Engineering
Swarnandhra College of
    Engineering and Technology
Narsapur, India

Palanisamy Sivasubramanian
Department of Mechanical
    Engineering
Dilkap Research Institute of
    Engineering and Management
    Studies
Raigad, India

Srinivasan V
Department of Manufacturing
    Engineering
Annamalai University
Chidambaram, India

Sahil Srivastava
Chandigarh University
Mohali, India

Alok Suna
Department of Mechanical
    Engineering
Indian Institute of Technology
    Bombay
Mumbai, India

Bunty Tomar
Laboratory for Advanced
  Manufacturing and Processing
  (LAMP)
Indian Institute of Technology
Jammu, India

Prameet Vats
Department of Mechanical
  Engineering
Indian Institute of Information
  Technology, Design and
  Manufacturing
Chennai, India

Alex Y
Laboratory for Advanced Research
  in Polymeric Materials
  (LARPM), School for Advanced
  Research in Polymers (SARP)
Central Institute of Petrochemicals
  Engineering & Technology
  (CIPET)
Bhubaneswar, India

Nitin Yadav
Department of Mechanical
  Engineering
Sant Longowal Institute of
  Engineering and Technology
Sangrur, India

Preface

As a popular method of material processing, additive manufacturing (AM) has been in use for two decades, but its practice as a significant commercial manufacturing technique has just lately begun to emerge. Nowadays, 3D printing is proven to be a cutting-edge technology for the revolutionized production of parts and devices. Hybrid AM is the term used to describe joining metal AM technology with conventional, subtractive technology, enabling each process to work together on the same machine and even on the same part. Hybrid manufacturing reduces the risks and costs associated with adopting metal AM technology, providing a more pragmatic and evolutionary pathway for industrial manufacturers.

The comprehensive subject matter is organized into 13 chapters: Chapter 1 discusses the introduction to hybrid metal AM technology. Chapters 2 and 3 present the role of wire-arc AM technology in modern manufacturing industries, and Chapter 4 discusses the joining of polymers to metals by hybrid AM techniques. Chapter 5 presents the heat treatments, microstructure, and mechanical properties of selective laser-melted AISI 316L. Chapters 6 and 7 deal with the advancements and challenges of hybrid AM composite materials. Chapters 8 and 9 present the role of AM technology for bio-based materials and plastic waste. Chapter 10 discusses the characterization of titanium feedstock powder prepared by recycling machining chips using the ball milling technique. The role of AM techniques in Industry 4.0 is presented in Chapter 11. Chapter 12 deals with the advancements and prospects of digital twin-driven AM technique. Finally, the concluding Chapter 13 deals with the role of AM in society.

The knowledge presented in this book is, all things considered, of significant value to scholars working in academia and industry who are interested in being updated in this field. Researchers, as well as practitioners in advanced manufacturing machines, engineers, and managers will find this book to be a valuable resource. This book can also serve as a textbook for students taking courses at the undergraduate and postgraduate levels.

Introduction

Over the past decade, additive manufacturing (AM) has become a powerful tool in manufacturing industries that converts digital model data into complex three-dimensional objects. AM technology builds the objects from the bottom to the top by fusing and binding materials layer by layer. AM technology offers the fabrication of customized objects out of metals, ceramics, and polymers without using molds or conventional manufacturing processes. However, researchers and companies are becoming more interested in metallic materials among these materials since these are commonly used engineering materials.

Metal AM technologies have had a revolutionary influence on the manufacturing sector in the past several decades. Metallic objects that could never have been manufactured merely a few years ago may now be manufactured with unprecedented strength and standards, utilizing a range of materials and metal AM technologies. Quality objects can be developed and built by layering metal powders with an energy source or a binding agent. Metal AM is widely employed in the aerospace, turbomachinery, biomedical, and other manufacturing industries that require lightweight and high strength-to-weight ratio products.

Despite several advantages of metal AM, some manufacturing industries do not consider this technology to be a viable manufacturing alternative since it fails to meet essential accuracy, geometrical tolerances, surface finish, and component accuracy criteria. Metal AM involves some post-processing since surface finishes and dimensional accuracy are sometimes inferior to traditional industrial operations. AM's layering and various interfaces might result in defects in the products. Thus, post-processing operations are required to resolve such issues.

Hybrid metal AM is one of the most effective approaches to overcome the shortcomings of these processes. As compared to a single manufacturing method, hybrid metal AM is a cutting-edge fabrication approach that combines more than two manufacturing processes (additive and subtractive) to manufacture more reliable, productive, and long-lasting components. The primary goal of this technology is to transform raw materials into finished products in a single machine or workstation. Hybrid metal

AM is an emerging and rapidly embraced technological marvel that has found its footing across diverse industries. This cutting-edge innovation holds immense potential, particularly in the realms of aerospace, medical devices, and automotive, where its applications have shown great promise.

This book aspires to illuminate the prospects of this avant-garde technology within industries and the educational landscape, with the aim of inspiring upcoming generations of researchers and engineers. To achieve this goal, a harmonious equilibrium is sought through the exploration of robust economic, environmental, and social criteria. By analyzing their interdependencies and applying these insights, this book provides invaluable guidance for driving technological innovation within the distinct frameworks of economy, environment, and society. This book offers an exceptional means for readers to grasp the intricacies of sustainable materials and manufacturing technologies with remarkable clarity and understanding. The book chapters cover topics such as recent advancements and challenges associated with hybrid metal AM, including the role of AM in sustainability, advances in metal AM technologies, sustainability solutions for metal AM processes, major opportunities, constraints, and economic considerations for design for metal AM, the importance of post-processing the metal AM, the role of metal AM in the era of Industry 4.0, etc.

Chapter 1

Hybrid metal additive manufacturing technology

Arun Kumar Bambam, Prameet Vats,
Alok Suna, and Kishor Kumar Gajrani
Indian Institute of Information Technology,
Design and Manufacturing

1.1 INTRODUCTION

Additive manufacturing (AM) is a fabrication technique in which parts are fabricated by adding material layer by layer on the basis of 3D model data. AM is applicable to a variety of materials, including polymers, metals, ceramics, and composites. There is a renewed emphasis on adopting AM to process metallic alloys for aerospace, automotive, and medical applications due to its design flexibility, higher degree of customization, and better material usage than conventional manufacturing [1,2].

1.1.1 Metal additive manufacturing processes

Metal AM processes are gaining a wide range of applications, such as aircraft, automobiles, medicine, and defense, as they can make complicated parts with high accuracy and quality [3]. Metal AM is creating new opportunities for designers and engineers to create innovative products and solutions by removing many of the limitations of conventional manufacturing techniques [4,5]. There are several direct and indirect metal AM processes, as shown in Figure 1.1. This section briefly discusses direct metal AM, and the key points are summarized in Table 1.1.

- *Powder bed fusion (PBF)*: This technology is able to generate high-resolution components and features with dimensional constraints and the ability to create parts with internal channels, offering two significant advantages over conventional manufacturing techniques. A PBF involves selectively melting a thin layer of metal powder with a high-powered source. The process is repeated layer by layer, resulting in a fully dense metal part [7]. PBF system is comprised of a confined chamber with an energy source (electron beam or laser), a powder bed, a scanner, a recoater arm or rake, a metal supply bed, and typically an inert gas supply, as shown in Figure 1.2. The most common techniques in PBF are selective laser melting (SLM) and electron beam melting (EBM).

DOI: 10.1201/9781003406488-1

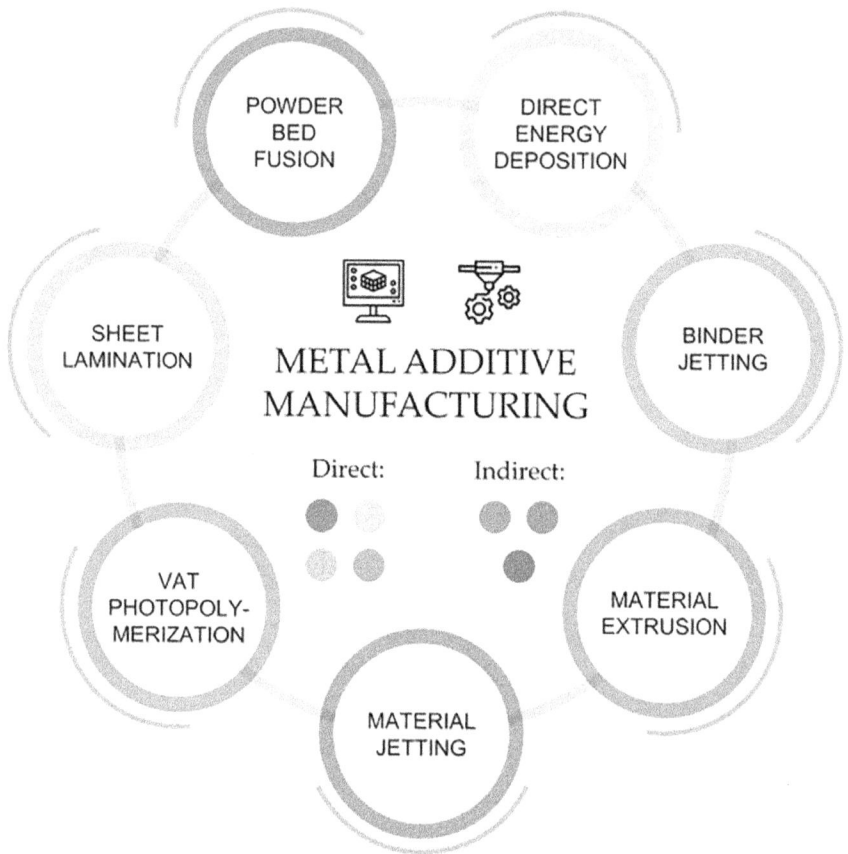

Figure 1.1 Direct and indirect metal additive manufacturing processes [6].

Table 1.1 Mechanism, sources, and materials form used in direct metal additive manufacturing processes

Processes	Mechanism	Source	Materials form
Powder bed fusion	Melting, sintering	Electron beam, laser	Powder
Directed energy deposition	Welding, cladding	Electron beam, laser	Powder, wire
Binder jetting	Binding	Binder	Powder
Sheet lamination	Ultrasonic welding, clamping and brazing	Laser, Computerized Numerical Control (CNC) cutter	Sheet

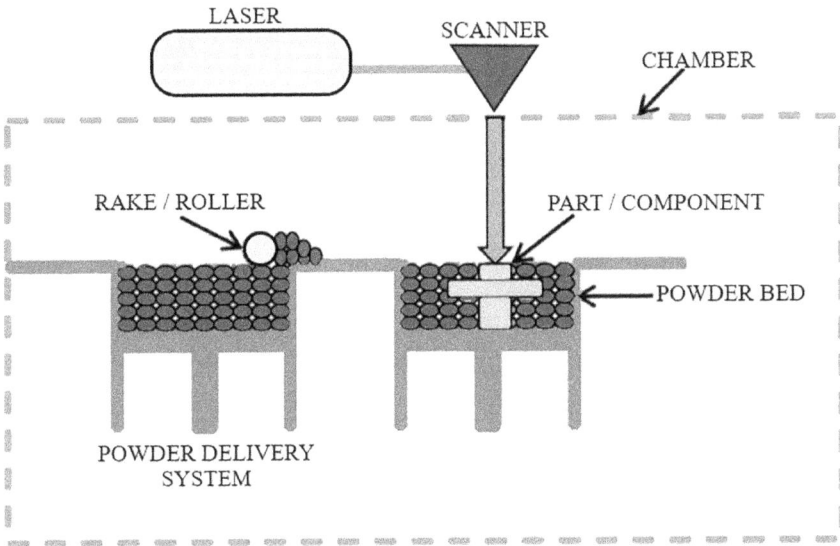

Figure 1.2 Schematic diagram of powder bed fusion.

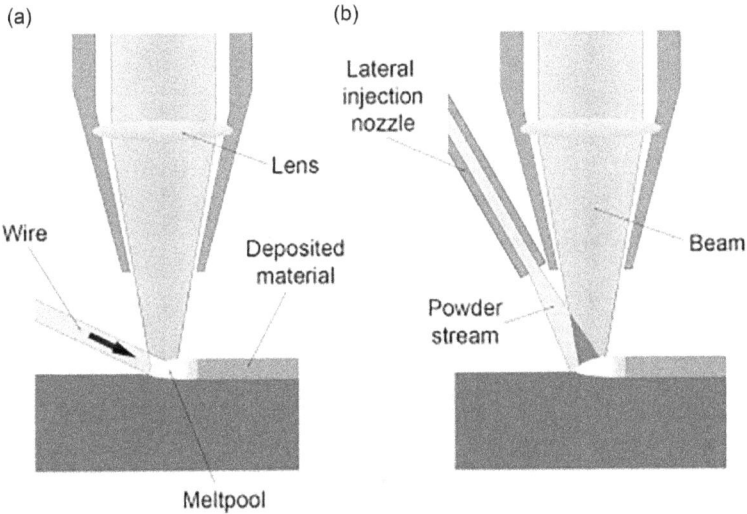

Figure 1.3 Schematic of wire and powder feed direct energy deposition [8].

- *Directed energy deposition (DED)*: This is a technique that involves feeding wire or metal powder into a focused energy source (electron beam or laser), as shown in Figure 1.3. The heat from the energy source melts the metal and fuses it to the existing layer or structure, forming a completely dense component. This technique is often used to add or repair components/parts by material deposition. This process is mostly popular due to the fact that it is substantially faster than other technologies such as PBF. Mechanical characteristics can also be tuned through managing the microstructure by chemical composition and the adjustment of the heat gradient [8,9].

- *Binder jetting (BJ)*: This is the process of depositing a liquid binder, layer by layer, onto a bed of metal powder, as shown in Figure 1.4. The binder binds the powder particles to form a green part, which is then sintered in a furnace to create a dense metal part. One of the biggest disadvantages of BJ fabricated parts is the comparatively low density (only 50%–60% in certain circumstances) [10,11].

- *Sheet lamination*: Sheet lamination or laminated object manufacturing (LOM) is the process of fusing metal sheets together with heat and pressure. To form a completely dense part, the sheets are cut into the required shape and joined together using a bonding agent or heat, as shown in Figure 1.5. It is a low-temperature technique used to create display models [12,13]. The technique is quite beneficial for quickly creating visual models at affordable/low prices. However, the approach is not ideal for creating structural or functional models. As a result, the sheet lamination procedure is only used for prototyping.

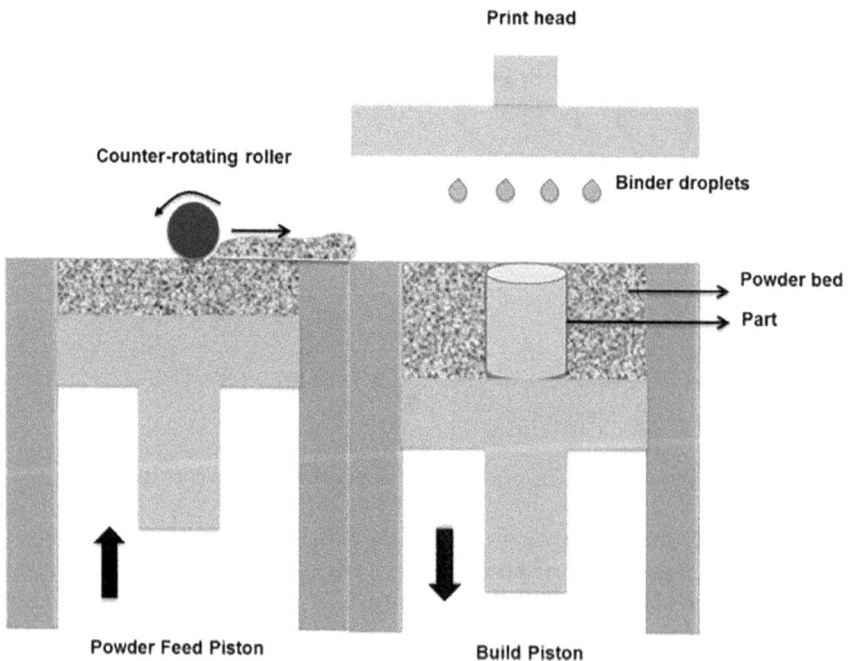

Figure 1.4 Schematic of binder jetting [10].

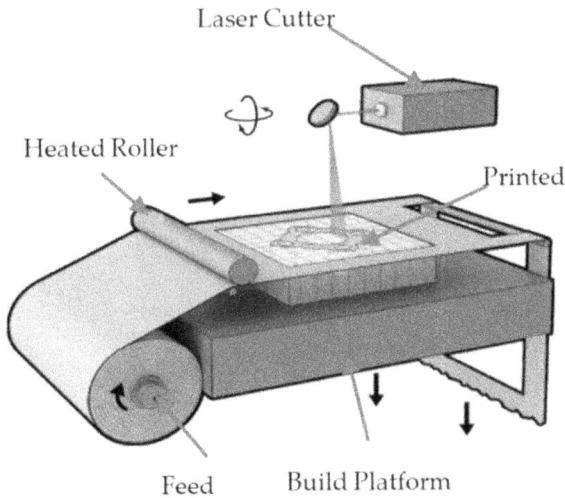

Figure 1.5 Schematic of sheet lamination [13].

AM has several advantages, including a broad variety of materials, a reduction in component weight, a lack of tooling and fixtures, and freeform fabrication. However, there are a few hurdles to direct adoption in metal AM, such as poor surface quality (i.e., rough and uneven surface profile) and inadequate dimensional accuracy in fabricated products, as well as very high investment costs for AM machines, material feedstock, and operating expenses [14,15]. To address these limitations, hybrid manufacturing methods (such as combinations of two or more manufacturing processes) can be utilized. The International Academy for Production Engineering defines hybrid manufacturing as "combining two or more established manufacturing techniques into a new integrated set-up wherein the benefits of each discrete process may be leveraged synergistically" [16,17].

1.1.2 Hybrid manufacturing

Hybrid manufacturing combines additive and subtractive manufacturing (SM) processes to create high-quality parts. The process involves using AM techniques to create the basic shape or geometry of the part and then using SM techniques to refine the shape and finish the surface, as shown in Figure 1.6. Hybrid manufacturing can also be used to create parts with a combination of materials [17]. For example, the AM process could be used to create a plastic part, while the SM process could be used to add metal components to the part. Overall, hybrid manufacturing is a promising approach to manufacturing that combines the strengths of multiple processes to create a more efficient and cost-effective manufacturing process, as shown in Figure 1.7. As technology continues to advance, it is likely that hybrid manufacturing will become increasingly common in a wide range of industries.

Design	Conversion	File Transfer	Configuration	Print
3D CAD file creation.	STL file conversion.	STL uploaded to slicing software.	Parameter optimization.	Parts printed are layer by layer.

Handover	Inspection	Heat Treatment	Machining	Removal
Parts are now finalized.	Examined for defects.	Tailors properties.	Improves surfaces and tolerances.	Parts are removed from the machine.

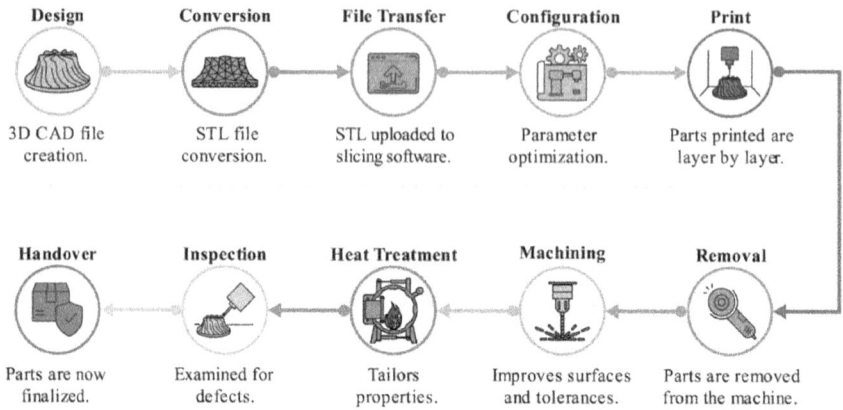

Figure 1.6 An example of a typical hybrid manufacturing process workflow [18].

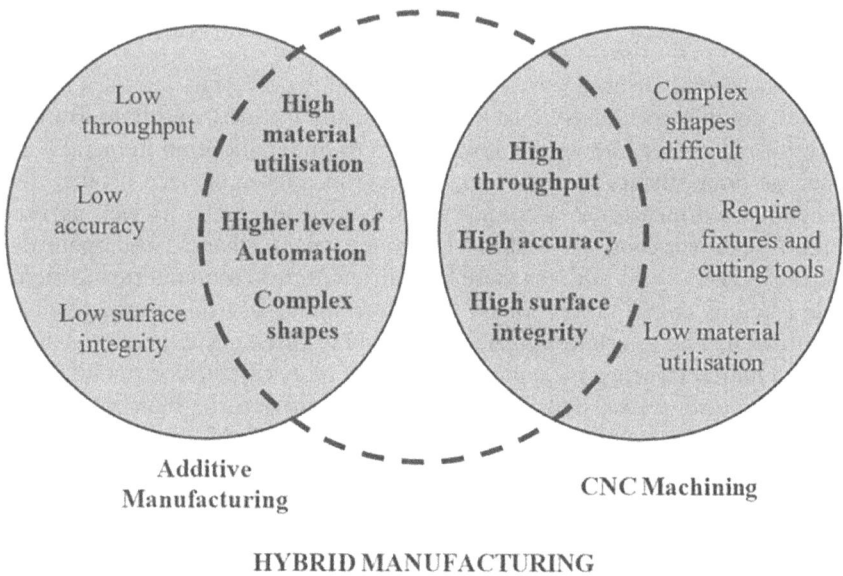

Additive Manufacturing

Low throughput
Low accuracy
Low surface integrity

High material utilisation
Higher level of Automation
Complex shapes

CNC Machining

High throughput
High accuracy
High surface integrity

Complex shapes difficult
Require fixtures and cutting tools
Low material utilisation

HYBRID MANUFACTURING

Figure 1.7 Advantages of hybrid manufacturing [19].

1.2 OVERVIEW OF PROCESS HYBRIDIZATION

Hybrid AM processes combine the benefits of two or more AM techniques or a combination of an AM technique with a subtractive or forming process. The hybridization of different manufacturing processes enables the production of complex parts with better accuracy, surface finish, and material properties than conventional AM or SM processes [17,20], as shown in Figure 1.8. Hybrid AM processes may be divided into three categories based on the combination of different manufacturing techniques, as shown in Figure 1.9.

	Properties	Machining	Additive Manufacturing	Hybrid Manufacturing
Design	Build Volume	L [======] H	L [==] H	L [====] H
	Geometric Complexity	L [=====] H	L [====] H	L [=====] H
	Design Versatility	☑ Limited ☐ Flexible	☐ Limited ☑ Flexible	☐ Limited ☑ Flexible
	Reshaping Products	☑ Rare ☐ Frequent	☑ Rare ☐ Frequent	☐ Rare ☑ Frequent
Materials	Material Waste	L [===] H	L [====] H	L [===] H
	Sustainability	L [==] H	L [=====] H	L [====] H
	Material Availability	☐ Limited ☑ Several	☑ Limited ☐ Flexible	☑ Limited ☐ Flexible
	Multi-material Usage	☑ Rare ☐ Frequent	☐ Rare ☑ Frequent	☐ Rare ☑ Frequent
Manufacturing	Production Speed	L [=====] H	L [=] H	L [=] H
	Post-processing Time	L [=====] H	L [====] H	L [==] H
	Customization	☐ Yes ☑ No	☑ Yes ☐ No	☑ Yes ☐ No
	Decenteralization	☐ Yes ☑ No	☑ Yes ☐ No	☑ Yes ☐ No
	Single-step Hybrid Objects	☐ Yes ☑ No	☑ Yes ☐ No	☑ Yes ☐ No
Product Quality	Shape Complexity	L [===] H	L [=====] H	L [====] H
	Strength	L [====] H	L [===] H	L [====] H
	Surface Finish/Quality	L [=====] H	L [==] H	L [==] H
	Dimensional Accuracy	L [=====] H	L [===] H	L [==] H
	Repeatability	L [====] H	L [==] H	L [==] H
Cost	Investment Cost	L [==] H	L [====] H	L [====] H
	Tooling Cost	L [====] H	L [=====] H	L [====] H
	Labor Cost	L [=====] H	L [====] H	L [=====] H
	Skill Requirement	L [====] H	L [==] H	L [====] H
	Mass Production	☐ Rare ☑ Frequent	☑ Rare ☐ Frequent	☑ Rare ☐ Frequent
	Batch Size	☐ Small ☑ Large	☑ Small ☐ Large	☑ Small ☐ Large

Figure 1.8 Overview of machining, additive manufacturing, and hybrid manufacturing.

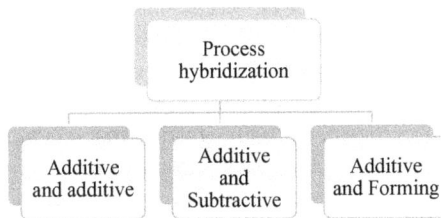

Figure 1.9 Probable categories of process hybridization.

1.2.1 Additive and additive

Additive and additive hybrid AM processes combine two or more AM techniques to produce parts with improved material properties, surface finish, and accuracy. For example, a combination of DED and PBF, or fused deposition modeling (FDM) and stereolithography (SLA) can be used to produce parts with high accuracy and surface finish.

1.2.2 Additive and subtractive

A subtractive process is combined with an AM approach in additive and subtractive hybrid AM processes. For example, combining SLM with milling can be used to create complicated components with excellent accuracy and surface finish. Similarly, PBF can be used to fabricate some cylindrical parts, and the surface finish can be improved using turning process.

1.2.3 Additive and forming

A forming process is combined with an AM technology in hybrid AM processes for better mechanical properties. For example, forging is coupled with DED to create components with good strength and better material properties.

1.3 HYBRID METAL-BASED ADDITIVE MANUFACTURING

Hybrid metal-based additive manufacturing (HMAM) is a new technique that combines the benefits of several AM technologies to create complex metal components more effectively and efficiently. In contrast to conventional SM techniques, which involve removing material from a solid block to mold into the desired shape, AM involves building up layers of material to produce the finished product [6,21]. HMAM combines many AM techniques, including PBF and DED, to create components with complex geometries, graded materials, and enhanced mechanical properties. One of the key benefits of HMAM is its potential to overcome the constraints of single AM techniques [6]. For example, PBF is good for precisely producing complex components, but it can be costly as well as time-consuming when producing large components. On the other hand, DED is excellent for swiftly assembling massive layers of metal, although it is less precise than PBF. Combining these two procedures allows HMAM to make big, complex components more accurately and quickly by taking advantage of each process's advantages.

The fabrication of components using graded materials, in which the material's composition progressively varies throughout the component, is possible with HMAM. This is especially helpful when producing components for industries such as aerospace or biomedical that need various material qualities [21,22]. For instance, a component should have a tougher material on the outside to fend against abrasion but a softer substance inside to absorb stress and lighten the component. HMAM may accomplish this by depositing various materials at various spots during the production process. In addition, HMAM has the promise of revolutionizing the manufacturing of metal parts with enhanced mechanical properties compared to conventional manufacturing techniques. HMAM may fabricate components with greater toughness, strength, and ductility by managing the deposition of metal powders and the energy input throughout the production process [6,23,24].

1.4 DESIGN FOR HYBRID METAL-BASED ADDITIVE MANUFACTURING PROCESSES

One of the primary benefits of HMAM systems is the ability to couple subtractive and additive processes to fabricate complex components. In order to ensure that the finished product satisfies the specified criteria, the part's design should take into consideration the prerequisites of both processes. For instance, the additive process could be more suitable if the component includes intricate details, but the subtractive method would be better suited if the component needs exact dimensional tolerances [25–28]. Therefore, while designing for HMAM, one can consider the following points:

- *Material selection*: The materials available for HMAM systems are restricted since not all materials can be used in both subtractive and additive processes. Materials must be chosen carefully so that they may be used with both processes and have the right mechanical properties for the component or application. The thermal qualities of the material should also be taken into account when choosing a material since certain materials may be challenging to process owing to high melting points or high thermal conductivity [29].
- *Part orientation and support structures*: The orientation of the component throughout the manufacturing process may have a considerable influence on the final component/part accuracy and quality. It is crucial to orient the component in such a manner that the requirement for support structures can be minimized because they may affect the surface quality, increase the fabrication time, and be difficult to remove. Furthermore, it has been observed that support structures may also have an effect on the accuracy of the component. Therefore, it is important to consider the part orientation during the design phase [30,31].
- *Surface finish and post-processing*: The final product's surface quality is an essential consideration for several applications, such as aeronautical or medical components. HMAM techniques may create components with excellent surface finishes, but further post-processing may be necessary to get the appropriate degree of texture or smoothness. To ensure the final outcome matches the required specifications, the product should be designed with post-processing requirements, such as polishing or machining [32].

1.5 HYBRID METAL ADDITIVE MANUFACTURING PROCESSES

HMAM processes combine additive and SM processes into a single system, providing the advantages of both technologies to create complex metal parts with high precision and accuracy. In this section, the different types of hybrid metal AM systems, their advantages, and their applications are discussed.

1.5.1 Hybrid manufacturing systems with laser-based additive manufacturing

These systems use laser-based AM technologies such as PBF and DED in combination with SM processes such as drilling or milling, along with some secondary processes such as shot peening and laser shock peening, to produce high-strength complex metal components. The integration of subtractive and additive processes allows creation of intricate features and high-precision surfaces that are difficult to achieve using only one process [33].

1.5.2 Hybrid manufacturing systems with binder jetting-based additive manufacturing

These systems use BJ AM technologies in combination with conventional manufacturing processes such as drilling or milling and secondary processes such as reheating, forming to produce complex metal components. The integration of subtractive and additive processes allows for the creation of parts with high resolution and accuracy. These systems are widely used in the medical and dental industries, where the creation of personalized and complex parts is required [34].

1.5.3 Hybrid manufacturing systems with sheet lamination-based additive manufacturing

These systems use sheet lamination-based AM processes coupled with conventional manufacturing processes such as drilling or milling to produce complex metal components. The combination of additive and subtractive practices allows for the creation of large and complex parts with high precision and accuracy. These systems are widely used in the energy and power generation industries, where the creation of large and complex components such as heat exchangers or turbines is required [35].

Apart from the above-mentioned hybridized process, researchers or engineers can develop many more possible combinations for their applications and facilitate industries with more new ideas, such as metal AM coupled with forming and many other secondary post-processing techniques.

1.6 BENEFITS OF HYBRID METAL-BASED ADDITIVE MANUFACTURING

Hybrid manufacturing, also known as hybrid additive-subtractive manufacturing or integrated manufacturing, integrates SM or AM processes for fabricating complex parts and structures. This technology has gained popularity due to its ability to bring together the advantages of both SM and AM, including decreased production time, increased design freedom, higher accuracy, and reduced waste [36–38]. This section covers the possible advantages of hybrid manufacturing.

- *Increased design freedom*: Hybrid manufacturing facilitates the fabrication of complex shapes that would be difficult to fabricate using conventional manufacturing processes. The combined use of SM and AM enables the fabrication of complex shapes and interior cavities that SM alone cannot achieve. This technology provides designers and engineers with new opportunities to build innovative products that have better functionality.
- *Reduced production time*: Hybrid manufacturing can substantially reduce manufacturing time by integrating the benefits of both SM and AM methods. Unlike conventional SM processes, which require multiple steps and different machinery and tools, hybrid manufacturing processes can be executed using a single machine. This may lead to a decrease in idle time, tool changing time, and work load time.
- *Improved accuracy*: Hybrid manufacturing facilitates the fabrication of components/parts with high precision and accuracy. The hybrid processes offer the fabrication of precise features with great reproducibility and precision, such as small channels and holes, which is not possible using single operations either by SM or AM processes.
- *Reduced waste*: Researchers are developing different strategies to reduce waste in manufacturing processes. Waste reduction plays an essential role in optimizing the total manufacturing cost. In this direction, hybrid manufacturing comes out as a potential process that minimizes overall material waste. The combination of SM and AM processes facilitates the manufacturing of complex geometries with less material waste than conventional SM methods. Researchers are also considering the total machining steps in designing steps to reduce the waste.
- *Cost-effective*: Hybrid manufacturing is a cost-effective technology that can reduce production costs and improve profitability. By combining the advantages of AM and SM processes, hybrid manufacturing can reduce the number of manufacturing steps, the number of machines required, and the amount of labor required.

1.7 ROLE OF HYBRID METAL ADDITIVE MANUFACTURING IN SUSTAINABILITY

Over time, the manufacturing sector and its facilities have accommodated the growing societal demand for goods and services while also having an effect on the ecosystem. This impact is characterized by the resources and energy consumption required for the inputs and outputs for the manufacturing processes and, on the other side, the consequences that our behavior has on the ecosystem (damage, diseases, resource depletion, biodiversity loss, changes in environments, etc.) [39,40]. Therefore, sustainability becomes a major field of interest in the researcher's community. Researchers are trying to contribute to sustainability in their own way. For example, Bambam et al. [41–43] focus more on environmentally

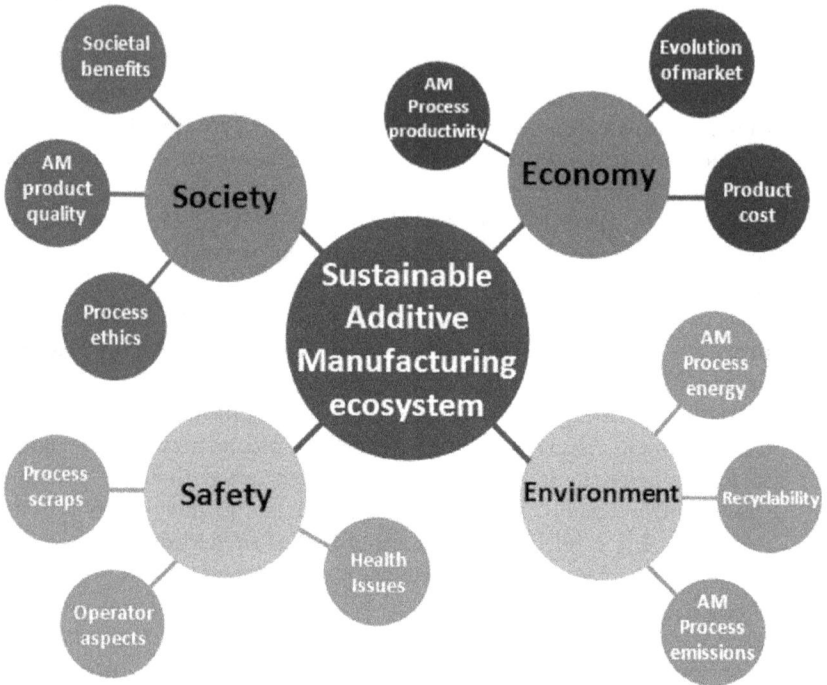

Figure 1.10 Sustainable additive manufacturing ecosystem [44].

friendly metalworking fluids so that they can reduce atmospheric emissions and minimize health risks for the workers. Furthermore, this research group added ionic liquids to vegetable oil to improve its properties so that friction, tool wear, and surface roughness can be minimized for SM applications. It also reduces tool cost, energy cost, and work piece rejection.

This section discusses the importance of hybrid manufacturing in sustainability, with an emphasis on its effects on social, economic, and environmental sustainability, as shown in Figure 1.10. There are several ways that hybrid manufacturing might help to advance sustainability. It is an environmentally friendly manufacturing method owing to its capacity to minimize material waste and preserve resources. Its flexibility and capacity to lower the costs of production contribute to economic sustainability and facilitate small enterprises to compete in the market [44,45]. Furthermore, it is a socially sustainable manufacturing method because of its capacity to enhance working conditions and support employment growth. Hybrid manufacturing has several benefits over conventional processes, which make it a potential approach for sustainable manufacturing.

1.7.1 Environmental sustainability

Hybrid manufacturing reduces material waste by using the precise quantity of material required to fabricate a part. This helps preserve energy and materials by minimizing waste generation. Unlike conventional manufacturing processes, which entail removing material from a larger sheet or block, there is less waste generation and more efficient use of available resources during production using hybrid manufacturing. Hybrid manufacturing further reduces environmental impact since it permits the use of sustainable resources such as bioplastics and recycled materials and considers some other steps to analyze sustainability as shown in Figure 1.11. It uses less energy and produces less waste, which reduces the carbon footprint. Therefore, fewer greenhouse gases are released into the atmosphere, which contributes to mitigating climate change [44,46].

1.7.2 Economic sustainability

Hybrid manufacturing provides financial advantages as well, most notably in the form of lower production costs. Large investments in equipment and machinery are necessary for traditional production processes, making it difficult for new entrants to break into the market. In contrast, hybrid production enables more production flexibility and reduces fixed manufacturing costs. This facilitates long-term economic growth by lowering entrance barriers and increasing competition for small enterprises. Hybrid production allows for more rapid prototyping and mass production, which in turn shortens product development cycles and raises quality standards. The time

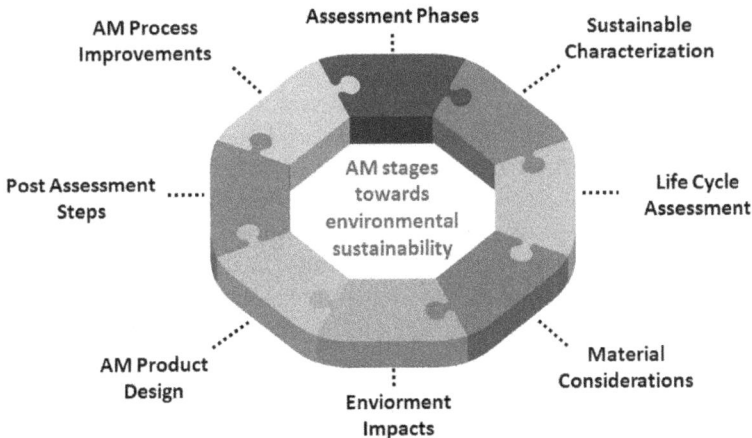

Figure 1.11 Additive manufacturing stages toward environmental sustainability [44].

and money needed to bring a product to market may be cut significantly by improving manufacturing efficiency through rapid prototyping and design iteration. Reduced expenses and higher profits are two outcomes of enhanced efficiency that may help the economy last [44,46].

1.7.3 Social sustainability

Hybrid manufacturing also provides societal advantages, including better working conditions for employees. Traditional methods for manufacturing products involve a lot of manual labor, and workers are often exposed to hazardous materials and situations. In contrast, hybrid manufacturing is highly automated and requires a minimal amount of human labor, which in turn promotes workers safety and improves productivity [44,47]. This can promote social sustainability and enhance work satisfaction. Hybrid manufacturing also facilitates customization, which is beneficial for customers demands as well as for society. Customers may personalize and customize items to their unique requirements and tastes, reducing waste and extending product life. Furthermore, small-batch production is made possible by hybrid production, opening the door to niche markets and customization—two factors that, in turn, may increase customer satisfaction and loyalty [48].

1.8 LIMITATIONS OF HYBRID METAL ADDITIVE MANUFACTURING PROCESSES

Hybrid AM processes have some limitations, such as skilled labor, high cost, complex processes, and limited material compatibility, that should be considered before adopting hybrid processes [1,16].

- *Complex process*: Hybrid AM processes are more complex than traditional AM or SM processes. The combination of different manufacturing processes requires specialized software and equipment to control the process, which increases the overall complexity.
- *High cost*: The cost of specialized software and equipment required for hybrid AM is very high, which limits the initial adoption of hybrid processes. Also, these processes required skilled labor, which demands high wages. Therefore, it's difficult for many small industries to adopt hybrid processes.
- *Limited material compatibility*: Hybrid AM processes are limited by the compatibility of the materials used in the process. The combination of different materials may result in material compatibility issues, limiting the range of materials that can be used in the process. Table 1.2 summarizes some other challenges associated with hybrid metal AM.

Table 1.2 Challenges associated with hybrid metal AM

Software integration	Hardware integration	Intelligent manufacturing

• CAM-hybrid simulation
• Real-time control and monitoring
• Tool path generation
  • Measuring and inspection tools
  • Hybrid tool path
  • Machining strategies
  • Material deposition strategies
	• System for remaining row material removal
	• System for metalworking fluids removal
	• Intermediate post-processes
	• High degree of freedom system such as robotic arms
		• Cloud manufacturing
		• IoT enable manufacturing

1.9 SUMMARY

In the past two decades, AM has emerged as a cutting-edge manufacturing technique that offers distinct benefits over conventional production methods. However, its industrial acceptance has not yet been achieved due to the limitations inflicted by the nature of the method. The combination of SM, AM, and other supplementary practices opens up new opportunities to overcome some of the major downsides of metal AM. As a result, hybrid metal AM has lately received interest from both academia and industry in contemporary manufacturing. Further advancements are projected to increase the availability of hybrid manufacturing systems in the future.

This chapter focuses on hybrid metal AM by discussing the hybridization of processes that are compatible with AM technologies, possible hybrid metal AM processes, and design considerations for hybrid AM technologies. Furthermore, the role of hybrid AM in sustainability is discussed. This chapter categorizes the essential information for engineers, professionals, and researchers working on the subject, as well as clearly highlighting the current difficulties and research possibilities.

REFERENCES

[1] Praveena, B. A., Lokesh, N., Buradi, A., Santhosh, N., Praveena, B. L., Vignesh, R. (2022). A comprehensive review of emerging additive manufacturing (3D printing technology): methods, materials, applications, challenges, trends and future potential. *Materials Today: Proceedings, 52*, 1309–1313.

[2] Horn, T. J., Harrysson, O. L. (2012). Overview of current additive manufacturing technologies and selected applications. *Science Progress, 95(3)*, 255282.

[3] Madhavadas, V., Srivastava, D., Chadha, U., Raj, S. A., Sultan, M. T. H., Shahar, F. S., Shah, A. U. M. (2022). A review on metal additive manufacturing for intricately shaped aerospace components. *CIRP Journal of Manufacturing Science and Technology, 39*, 18–36.

[4] Frazier, W. E. (2014). Metal additive manufacturing: a review. *Journal of Materials Engineering and Performance*, 23, 1917–1928.

[5] Bandyopadhyay, A., Zhang, Y., Bose, S. (2020). Recent developments in metal additive manufacturing. *Current Opinion in Chemical Engineering*, 28, 96–104.

[6] Pragana, J. P. M., Sampaio, R. F., Bragança, I. M. F., Silva, C. M. A., Martins, P. A. F. (2021). Hybrid metal additive manufacturing: a state-of-the-art review. *Advances in Industrial and Manufacturing Engineering*, 2, 100032.

[7] Frazier, W. E. (2014). Metal additive manufacturing: a review. *Journal of Materials Engineering and Performance*, 23, 1917–1928.

[8] Molitch-Hou, M. (2018). Overview of additive manufacturing process. In *Additive manufacturing* (p. 138). Oxford: Butterworth-Heinemann.

[9] Saboori, A., Aversa, A., Marchese, G., Biamino, S., Lombardi, M., Fino, P. (2019). Application of directed energy deposition-based additive manufacturing in repair. *Applied Sciences*, 9(16), 3316.

[10] Nandwana, P., Elliott, A. M., Siddel, D., Merriman, A., Peter, W. H., Babu, S. S. (2017). Powder bed binder jet 3D printing of Inconel 718: densification, microstructural evolution and challenges. *Current Opinion in Solid State and Materials Science*, 21(4), 207–218.

[11] Li, M., Du, W., Elwany, A., Pei, Z., Ma, C. (2020). Metal binder jetting additive manufacturing: a literature review. *Journal of Manufacturing Science and Engineering*, 142(9), 090801.

[12] Zhang, Y., Wu, L., Guo, X., Kane, S., Deng, Y., Jung, Y. G., Lee, J.H, Zhang, J. (2018). Additive manufacturing of metallic materials: a review. *Journal of Materials Engineering and Performance*, 27, 1–13.

[13] Vafadar, A., Guzzomi, F., Rassau, A., Hayward, K. (2021). Advances in metal additive manufacturing: a review of common processes, industrial applications, and current challenges. *Applied Sciences*, 11(3), 12–13.

[14] Maleki, E., Bagherifard, S., Bandini, M., Guagliano, M. (2021). Surface post-treatments for metal additive manufacturing: progress, challenges, and opportunities. *Additive Manufacturing*, 37, 101–619.

[15] Abdulhameed, O., Al-Ahmari, A., Ameen, W., Mian, S. H. (2019). Additive manufacturing: challenges, trends, and applications. *Advances in Mechanical Engineering*, 11(2), 1–27. https://doi.org/10.1177/1687814018822880.

[16] Altıparmak, S. C., Yardley, V. A., Shi, Z., Lin, J. (2021). Challenges in additive manufacturing of high-strength aluminium alloys and current developments in hybrid additive manufacturing. *International Journal of Lightweight Materials and Manufacture*, 4(2), 246–261.

[17] Dilberoglu, U. M., Gharehpapagh, B., Yaman, U., Dolen, M. (2021). Current trends and research opportunities in hybrid additive manufacturing. *The International Journal of Advanced Manufacturing Technology*, 113, 623–648.

[18] Armstrong, M., Mehrabi, H., Naveed, N. (2022). An overview of modern metal additive manufacturing technology. *Journal of Manufacturing Processes*, 84, 1001–1029.

[19] Jiménez, A., Bidare, P., Hassanin, H., Tarlochan, F., Dimov, S., Essa, K. (2021). Powder-based laser hybrid additive manufacturing of metals: a review. *The International Journal of Advanced Manufacturing Technology*, 114, 63–96.

[20] Schneberger, J. H., Kaspar, J., Vielhaber, M. (2020). Post-processing and testing-oriented design for additive manufacturing-a general framework for the development of hybrid AM parts. *Procedia CIRP*, 90, 91–96.

[21] Korkmaz, M. E., Waqar, S., Garcia-Collado, A., Gupta, M. K., Krolczyk, G. M. (2022). A technical overview of metallic parts in hybrid additive manufacturing industry. *Journal of Materials Research and Technology*, 18, 384–395.

[22] Stavropoulos, P., Bikas, H., Avram, O., Valente, A., Chryssolouris, G. (2020). Hybrid subtractive-additive manufacturing processes for high value-added metal components. *The International Journal of Advanced Manufacturing Technology*, 111, 645–655.

[23] Kumar, M. B., Sathiya, P. (2021). Methods and materials for additive manufacturing: a critical review on advancements and challenges. *Thin-Walled Structures*, 159, 107–228.

[24] Tepylo, N., Huang, X., Patnaik, P. C. (2019). Laser-based additive manufacturing technologies for aerospace applications. *Advanced Engineering Materials*, 21(11), 1900617.

[25] Lin, X., Zhu, K., Fuh, J. Y. H., Duan, X. (2022). Metal-based additive manufacturing condition monitoring methods: from measurement to control. *ISA Transactions*, 120, 147–166.

[26] Webster, S., Lin, H., Carter III, F. M., Ehmann, K., Cao, J. (2021). Physical mechanisms in hybrid additive manufacturing: a process design framework. *Journal of Materials Processing Technology*, 291, 117048.

[27] Thompson, M. K., Moroni, G., Vaneker, T., Fadel, G., Campbell, R. I., Gibson, I., Bernard, A., Schulz, J., Graf, P., Ahuja, B., Martina, F. (2016). Design for additive manufacturing: trends, opportunities, considerations, and constraints. *CIRP Annals*, 65(2), 737–760.

[28] Vaneker, T., Bernard, A., Moroni, G., Gibson, I., Zhang, Y. (2020). Design for additive manufacturing: framework and methodology. *CIRP Annals*, 69(2), 578–599.

[29] Malaga, A. K., Agrawal, R., Wankhede, V. A. (2022). Material selection for metal additive manufacturing process. *Materials Today: Proceedings*, 66, 1744–1749.

[30] Stavropoulos, P., Foteinopoulos, P., Papacharalampopoulos, A., Bikas, H. (2018). Addressing the challenges for the industrial application of additive manufacturing: towards a hybrid solution. *International Journal of Lightweight Materials and Manufacture*, 1(3), 157–168.

[31] Zhang, Y., Yang, S., Zhao, Y. F. (2020). Manufacturability analysis of metal laser-based powder bed fusion additive manufacturing: a survey. *The International Journal of Advanced Manufacturing Technology*, 110, 57–78.

[32] Ye, C., Zhang, C., Zhao, J., Dong, Y. (2021). Effects of post-processing on the surface finish, porosity, residual stresses, and fatigue performance of additive manufactured metals: a review. *Journal of Materials Engineering and Performance*, 30, 6407–6425.

[33] Sealy, M. P., Madireddy, G., Williams, R. E., Rao, P., Toursangsaraki, M. (2018). Hybrid processes in additive manufacturing. *Journal of manufacturing Science and Engineering*, 140(6), 060801.

[34] Gawali, S. K., Pandey, G. C., Bajpai, A., Jain, P. K. (2022). Large-part manufacturing using CNC-assisted material extrusion-based additive manufacturing: issues and challenges. *International Journal on Interactive Design and Manufacturing*, 17, 1185–1197.

[35] Bhatt, P. M., Kabir, A. M., Peralta, M., Bruck, H. A.,Gupta, S. K. (2019). A robotic cell for performing sheet lamination-based additive manufacturing. *Additive Manufacturing*, 27, 278–289.

[36] Ngo, T. D., Kashani, A., Imbalzano, G., Nguyen, K. T., Hui, D. (2018). Additive manufacturing (3D printing): a review of materials, methods, applications and challenges. *Composites Part B: Engineering, 143*, 172–196.

[37] Häfele, T., Schneberger, J. H., Kaspar, J., Vielhaber, M., Griebsch, J. (2019). Hybrid additive manufacturing-process chain correlations and impacts. *Procedia CIRP, 84*, 328–334.

[38] Xie, Y., Zhang, H., Zhou, F. (2016). Improvement in geometrical accuracy and mechanical property for arc-based additive manufacturing using metamorphic rolling mechanism. *Journal of Manufacturing Science and Engineering, 138*(11), 111002.

[39] Bambam, A. K., Gajrani, K. K. (2023). Challenges in achieving sustainability during manufacturing. In Navneet Khanna, Kishor Kumar Gajrani, Khaled Giasin, J. Paulo Davim (eds.), *Sustainable Materials and Manufacturing Technologies* (pp. 108–124). Boca Raton, FL: CRC Press. ISBN: 9781003291961. https://doi.org/10.1201/9781003291961-9.

[40] Álvarez, M. E. P., Bárcena, M. M., González, F. A. (2017). On the sustainability of machining processes. Proposal for a unified framework through the triple bottom-line from an understanding review. *Journal of Cleaner Production, 142*, 3890–3904.

[41] Bambam, A. K., Dhanola, A., Gajrani, K. K. (2023). A critical review on halogen-free ionic liquids as potential metalworking fluid additives. *Journal of Molecular Liquids, 380*, 121727.

[42] Bambam, A. K., Dhanola, A., Gajrani, K. K. (2022). Machining of titanium alloys using phosphonium-based halogen-free ionic liquid as lubricant additives. *Industrial Lubrication and Tribology, 74*(6), 722–728.

[43] Bambam, A. K., Alok, A., Rajak, A., Gajrani, K. K. (2022). Tribological performance of phosphonium-based halogen-free ionic liquids as lubricant additives. *Proceedings of the Institution of Mechanical Engineers, Part J: Journal of Engineering Tribology, 237*(4), 881–893.

[44] Javaid, M., Haleem, A., Singh, R. P., Suman, R., Rab, S. (2021). Role of additive manufacturing applications towards environmental sustainability. *Advanced Industrial and Engineering Polymer Research, 4*(4), 312–322.

[45] Ford, S., Despeisse, M. (2016). Additive manufacturing and sustainability: an exploratory study of the advantages and challenges. *Journal of Cleaner Production, 137*, 1573–1587.

[46] Devarajan, B., Bhuvaneswari, V., Arulmurugan, B., Narayana, A. V. N. S. L., Priya, A. K., Abbaraju, V. D. N., Mukunthan, K. S., Sharma, A. K., Ting, S. S., Masi, C.(2022). Hybrid novel additive manufacturing for sustainable usage of waste. *Journal of Nanomaterials, 2022*, 12. https://doi.org/10.1155/2022/2697036.

[47] Chandra, M., Shahab, F., KEK, V., Rajak, S. (2022). Selection for additive manufacturing using hybrid MCDM technique considering sustainable concepts. *Rapid Prototyping Journal, 28*(7), 1297–1311.

[48] Agnusdei, L., Del Prete, A. (2022). Additive manufacturing for sustainability: a systematic literature review. *Sustainable Futures, 4*, 100098.

Chapter 2

An overview of wire arc additive manufacturing (WAAM) technique with different alloys in modern manufacturing industries

Tuhina Goswami
KGEC

Shatarupa Biswas
NIT, Silchar

Santanu Das
KGEC

Manidipto Mukherjee
CSIR-CMERI

2.1 INTRODUCTION

Additive manufacturing (AM) is a rapid evolution process that is widely accepted by the manufacturing sector for its capability of large-scale component production at a quick rate of deposition. AM is the process of combining materials to produce items from 3D model data, generally continuous layer deposition on a layer, according to the American Society for Testing and Materials (ASTM) [1]. Based on ASTM standards, four categories can be used to group all metal AM processes, such as powder bed fusion, lamination of the sheet, and binder jetting [1]. However, stereolithography (SL) is the initial application of AM which was developed in the year of 1987. The method is three-dimensional systems that can be used by a laser with UV light-sensitive liquid polymer to thin layers [1]. For developing a new method for 3D-building/printing items utilizing step-by-step deposition, AM technology has emerged as the major engine of the fourth industrial upheaval and a common industrial method. It has evolved into a cutting-edge and robust production methodology that generates the component from the CAD model. Also, complex component manufacturing can be done in a single step without the need for costly fixtures or assembly. AM technology is widely used in

DOI: 10.1201/9781003406488-2

engineering applications, such as the automotive, aviation, energy, and health sectors [2]. Usually, columnar grain structures produced by AM methods have grains that span several re-melt layers. Various types of AM methods are available in manufacturing industries, such as laser additive manufacturing, cold spray deposition, and WAAM [3].

2.2 WIRE ARC ADDITIVE MANUFACTURING (WAAM)

The wire arc additive manufacturing (WAAM) process uses the deposition of a layer-on-layer of metal wire to manufacture complex components. Moreover, WAAM provides advantages over powder or wire-based AM and laser or electron beam; its additional benefits include little material waste, high equipment adaptability, cheap running costs, fast production cycles, and outstanding forming quality. According to ISO/ASTM 52900 [1], WAAM is an energy deposition procedure; it also supplies and controls the wire deposition rate during the welding process. As a result, the final product is a fully dense component with 0.2 mm of dimensional accuracy [1]. However, WAAM may be created with thin or thick walls, solid construction, or casting without the need for dies and with improved mechanical qualities and microstructure. Nowadays, the WAAM method is used in a broad range of industries, including aircraft, automotive, plant and refinery, defense, and construction. Figure 2.1 shows the deposition process of WAAM techniques.

2.3 WELDING PROCESS USED IN WAAM

Arc-based welding methods are being successfully used by researchers in the WAAM technique. Gas tungsten arc welding (GTAW), gas metal arc welding (GMAW), and plasma arc welding (PAW) are the well-known welding methods that are basically used in the WAAM process, which underwent a

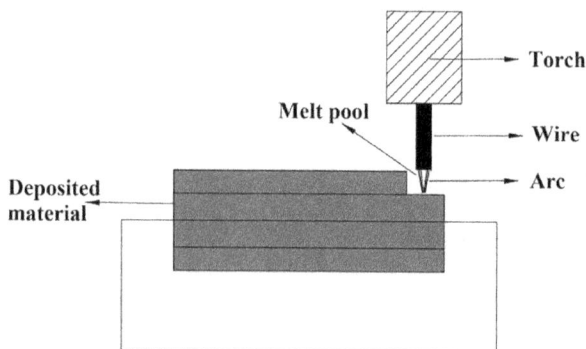

Figure 2.1 The process of WAAM techniques

proper solution when the cold metal transfer welding (CMT) process was substituted for GMAW. However, the CMT process can be improved with a minimum input of heat, spatter disruptions, and a high deposition rate [1,2]. The GTAW techniques are elaborated on in the following sections.

2.3.1 Gas tungsten arc welding (GTAW)

Among the various WAAM techniques, GTAW is the most popular. The GTAW technique has the advantage of allowing for the separation of the energy input and filler in certain ranges, as well as the use of many wires simultaneously. Because the wire feeder must alter its working position, the independent wire feed causes problems while changing welding directions. Additionally, the input variable and the use of a pulsed welding process might affect the form of the melt pool when employing the GTAW welding process [5]. Figure 2.2 shows the process of GTAW.

2.3.2 Gas metal arc welding (GMAW)

Another popular welding technique is GMAW, which has been utilized for surface cladding and AM processes for many years. The welding process, such as controlling short arc and spray arc processes, has been improved recently by the development of new control systems for arc welding power sources. Han et al. demonstrated that the material deposition rate for the AM process using a modified GMAW process may reach 4.71 kg/h [6]. Applying multi-wire techniques is another way to increase the bulk of the

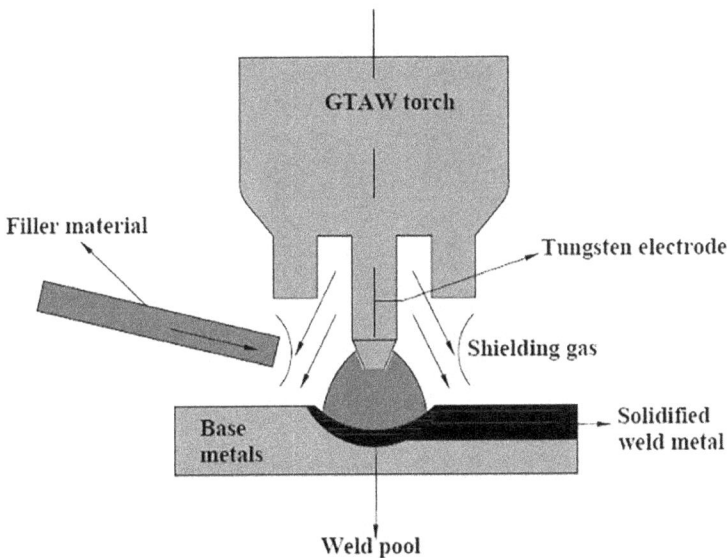

Figure 2.2 The process of GTAW techniques.

deposition of material. The mobility of the wire tip and the minimal geo-metrical error can be achieved by the AM process with the help of the GMAW (Figure 2.3) technique [6].

2.3.3 Plasma arc welding (PAW)

The PAW technique (Figure 2.4) is another welding process that technology has been using for decades to perform multi-layer surfacing or cladding, a very straightforward AM technique. Similar to tungsten inert gas welding, this procedure is also capable of separating the energy source and the material flow from the filler, and it may be used with a variety of materials (such as steel and co-base alloys). The droplet transfer model, like in other welding processes, is

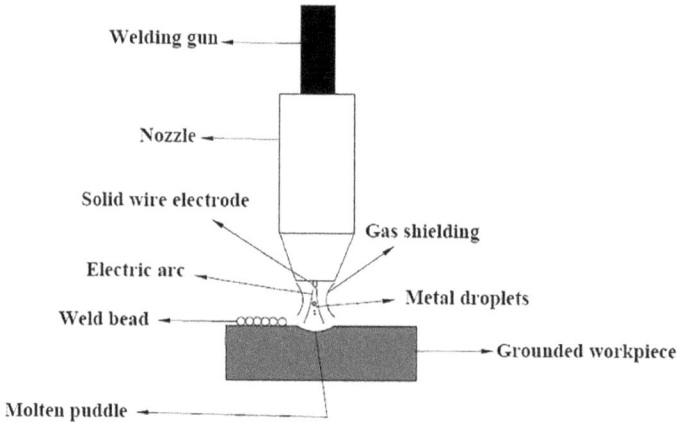

Figure 2.3 The process of GMAW techniques.

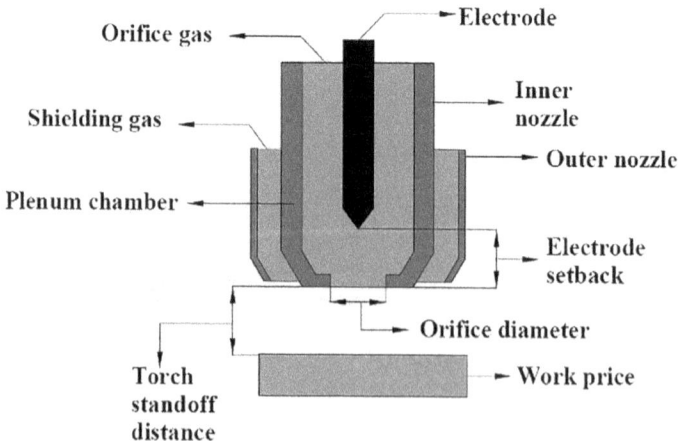

Figure 2.4 The process of PAW techniques.

affected by input variables and can be split into three distinct modes of transition based on the frequency and stability of droplet transfer [7-9].

2.3.4 Steps of the WAAM process

Various steps of the WAAM process are described in this current section. The first step is 3D representation using CAD or reverse engineering methods. Then, after 3D scanning and saving the model in a stl file format, the geometry of the model (3D to 2D layer) was sliced using a slicing software. After the software slices the model (3D part), the result shows the best parameters for material deposition and also creates the optimal path [1–4]. Next, to fabricate the components, it is essential to choose the proper input parameters, such as material feed rate (FS), table speed, arc current (I), gas flow rate, route strategy, and heat input. After that, the code is prepared by CNC programming. The first deposition is made, followed by the nozzle-specific height and formation of the next layer over the previous layer deposited [2].

2.4 MATERIALS USED IN WAAM TECHNIQUES

Many spooled wires are used in the welding industry for the WAAM process.

2.4.1 Titanium alloys

Due to its high strength-to-weight ratio, the titanium (Ti) alloy is used in the WAAM process. The deposition rate of Ti alloys is within 0.75–2 kg/h [2]. As a result, the metal layers that are produced are quite dense and do not require the Hot Isostatic Pressing (HIP) process. The size of the component is controlled by the range of the manipulator. Ti alloy deposited by WAAM has superior tolerance; however, it is better than wrought iron alloy under heavy cyclic stress. [2,10–12].

2.4.2 Aluminum alloys

Numerous fabrication trials have been successfully done for a variety of aluminum (Al) alloys, Al-Cu, Al-Si (4xxx), and Al-Mg (5xxx) alloys. Since the cost of traditional machining is relatively inexpensive for straightforward and small components, the WAAM technology is acceptable for complex parts with big and tinny walls. However, deposited Al alloy parts' mechanical qualities require post-process heat treatment to increase strength, improve mechanical properties, and enhance the microstructure [1,13–15].

2.4.3 Nickel-based superalloy

Superalloys based on nickel can maintain high strength at high temperatures. This material has found tremendous commercial applicability with

WAAM. Inconel 625 has been the primary alloy taken into consideration, and its qualities are examined [1,15,16].

2.4.4 Stainless steel

Stainless steel (SS) can also be used as a wire material in the WAAM process [2,17,18]. It has good mechanical characteristics and microstructure. Along with austenite forms, rapid cooling can also cause the production of non-equilibrium nitrides side by side. SS components have a small amount of anisotropy, but with the right post-process heat treatment and qualities, equivalent products can be made.

2.5 INPUT VARIABLES

It is essential to administer and control a wide range of input variables to produce stable and defect-free components. The important variables are described in the following sections.

2.5.1 Wire feed speed

The variation in bead height and wire FS needs to be linear. Therefore, a high-wire FS will produce a tall and narrow bead. The wetting angle and aspect ratio both decrease as wire feed speed (WFS) is raised. It is found that the roughness and melt across depth are independent of the WFS [2,19].

2.5.2 Speed of travel

The width of the bead, angle of wetting, and melting through depth all decrease with increasing speed of travel (TS), but the bead height does not change. According to the concept, melt-through depth will decrease as transit speeds increase and heat input decreases. Due to increased TS, there is a slight increase in bead roughness [2,19].

2.5.3 Arc current (I)

The bead roughness is the aspect of current (I) that is most significantly impacted, as bead roughness decreases with increasing, I value. As the I is increased, other responses such as the width of the bead, angle of wetting, and melting through depth will be raised [2,19].

2.6 LITERATURE REVIEW

Various research has been done on the WAAM process to optimize the input parameter and establish a relationship between parameters using various materials, which are shown in Table 2.1.

Table 2.1 Literature review on the WAAM process

Author	Year	Material	Input variables	Output variables	Technique used	Findings
Zhang et al. [13]	2013	Al alloy, AA6061			Laser-cold metal transfer	An optical microscope, SEM, and EDAX were used to analyze the microstructure. To evaluate the mechanical characteristics of the welded junction, tests were performed on the cross- weld tensile strength and hardness. The result was accepted joints with a finer microstructure and no flaws.
Qiu et al. [11]	2015	Ti alloy	Three-beam laser nozzle	Layer height	Direct laser deposition (DLD)	Large structures could be manufactured using the DLD technique produced samples exhibited high tensile strengths but poor elongations and were dominated by columnar grains and martensite. Due to insufficient remelting of the prior layer, planar holes were present, which significantly reduced ductility.
Xiong et al. [10]	2016	Ti alloy	I, voltage, heat dissipation condition, inter-layer temperature	Layer width	GMAW, MATLAB	A second-order Hammerstein model was used to determine the relationship between the width of the layer and TS in the nonlinear WAAM process. For an intriguing variable width control in several layers, an intelligent controller was created. Less than 0.5 5 mm was the largest absolute difference between the detected and predicted widths.
Wang et al. [16]	2016	Inconel 625			GTAW–ALD	In this study, the location shows the effect on the material characteristics of manufactured components. Weld pool behavior and temperature field measurement provides provide additional insight into the variance of microstructure and mechanical properties in different locations.

(Continued)

Table 2.1 (Continued) Literature review on the WAAM process

Author	Year	Material	Input variables	Output variables	Technique used	Findings
Campitelli et al. [14]	2020	Al alloy	I, Voltage, TS, Bead Volume, Electric energy	SLM (full build, single part), WAAM CMT mix drive	CMT, CMT mix drive	CMT and CMT mix drive, two different deposition processes for WAAM, were compared in an analysis used to create Al alloy-alloy components. Material characteristics were examined in two orthogonal directions: parallel direction and transverse direction. About the studied output, no significant changes were seen with different orientations.
Le and Mai [17]	2020	308L			GMAW- AM	Investigations were made into the characteristics of GMAW-AM. Columnar dendrites have commonly defined the microstructure of GMAW-AM at the bottom and middle part parts of the deposited wall. Equiaxed dendrites dominate the top portion of the wall.
Raut and Taiwade [1]	2021	Ni-Based alloys, Ti alloy, Al alloy, Steels,	I, TS, WFS		CMT	There has been an analysis of numerous WAAM-related information addressed. Included in these parameter optimizations, are process selection, process advancements, and post-process heat treatment to achieve microstructural homogeneity and dimensional accuracy.
Kumar and Manikand an [15]	2021	Al alloys, Ni-Based Superalloy s, Inconel 625			Finite eliminate method (FEA)	FEA simulation of WAAM processes is reviewed. The uneven temperature distribution was seen throughout the deposition process, which leads to increased residual stress and deformation. Additionally, the temperature decreases with deposition height and is lower in the upper than the lower section. The distribution of temperature is significantly impacted by the process parameters. Higher compressive and tensile residual stresses were observed in the thickness direction and base plate.

(Continued)

Table 2.1 (Continued) Literature review on the WAAM process

Author	Year	Material	Input variables	Output variables	Technique used	Findings
Huang et al. [19]	2021	Photoiniti ators, UV absorbers			3D Printing	All of the models used in this study had layers that were 0.02 02 mm thick. At room temperature and with outside air pressure, all models were printed, base plate and compressive in the thickness direction and longitudinal direction.
Nagasai et al. [18]	2022	308 308L, SS	WFS, I, Voltage, Rotational speed, Arc length correction	Average width of wall and height of the single layer, Dia of the cylinder, Height of cylindrical component, Number deposited layers	GMAW, CMT	The 308 308L SS cylindrical components can be produced using the GMAW and CMT processes with high density and few flaws. In both parts, the deposited layers are visible and devoid of significant holes and fissures. The failure modes are ductile fractures, and the manufactured components have excellent elongation. The fracture surfaces displayed huge dimples with uniformly distributed distributions.
Chakrabor ty et al. 2022 [12]	2022	Ti alloy				The WAAM technology is helpful for producing medium-to large-scale aerospace parts with higher precision in lesser time. The many WAAM aspects of various Ti alloys, their processing parameters, energy sources, and various postprocessing techniques are explained in this work.

2.7 CONCLUSIONS

From the aforementioned investigation, the following conclusions can be drawn:

Among the various materials, Ti alloy, SS, and Al alloy are commonly used in the WAAM technique. These metals provide better strength.

Various input parameters, such as WFS, substrate temperature, and TS, are chosen to complete the WAAM technique. However, the main output parameters are chosen such as layer width and layer height. WAAM technology is particularly helpful for producing large-scale aerospace parts with higher accuracy in less time. To know the mechanical characteristics of the welded junction, tensile strength and hardness are required. Higher compressive and tensile residual stresses were observed in the direction of deposition on the base plate.

2.8 FUTURE SCOPES

- Researchers can use other hard materials and choose other input and output parameters. Also, the parameters can be optimized with various optimizing tools to find the best parameters.
- Researchers can use various software programs to generate mathematical models to analyze stress.

REFERENCES

[1] Mohd N, Solomon DG, Bahari MF (2007) A review on current research trends in electrical discharge machining (EDM). *International Journal of Machine Tools & Manufacture.* 47(7–8): 1214–1228.

[2] DiBitonto DD (1969) Theoretical models of the electrical discharge machining process. I. A simple cathode erosion model. *Journal of Applied Physics.* 66(9): 4095–4103.

[3] Das S, Paul S, Doloi B (2020) Feasibility assessment of some alternative dielectric mediums for sustainable electrical discharge machining: a review work. *Journal of the Brazilian Society of Mechanical Sciences and Engineering.* 42(148): 1–21.

[4] Jain VK, Sidpara A, Balasubramaniam R, Lodha GS, Dhamgaye VP, Shukla R (2014) Micromanufacturing: a review - Part I. *Proceedings of the Institution of Mechanical Engineers, Part B: Journal of Engineering Manufacture.* 228(9): 973–994.

[5] Richard J, Demellayer R (2013) Micro-EDM-milling development of new machining technology for micro-machining. *Procedia CIRP.* 6: 292–296.

[6] Uhlmann E, Perfilov I (2018) Machine tool and technology for manufacturing of micro- structures by micro dry electrical discharge milling. *Procedia CIRP.* 68: 825–830.

[7] Gotoh H, Tani T, Okada M, Goto A, Masuzawa T, Mohri N (2013) Wire electrical discharge milling using a wire guide with reciprocating rotation. *Procedia CIRP.* 6: 199–202.

[8] Wang J, Qian J, Ferraris E, Reynaerts D (2016) Precision micro-EDM milling of 3D cavities by incorporating in-situ pulse monitoring. *Procedia CIRP.* 42: 656–661.

[9] Schulze V, Ruhs C (2010) On-machine measurement for the micro-EDM-milling process using a confocal white light sensor. Proceedings of the 10th International Conference of the European Society for Precision Engineering. Nanotechnology, EUSPEN 2: 37–40.

[10] Chakraborty S, Dey V, Ghosh SK (2015) A review on the use of dielectric fluids and their effects in EDM characteristics. *Precision Engineering.* 40: 1–6.

[11] Valaki JB, Rathod PP, Sidpara AM (2019) Sustainability issues in electric discharge machining. In: Gupta K (ed) *Innovations in manufacturing for sustainability. Materials forming, machining and tribology.* Springer, Cham.

[12] Das S, Paul S, Doloi B (2019) Investigation of the machining performance of neem oil as a dielectric medium of EDM: a sustainable approach. In: *IOP conference series: materials science and engineering*, IOP Publishing, KIIT Bhubaneswar, India. 653 (1): 1–8.

[13] Valaki JB, Rathod PP, Sankhavara CD (2016) Investigations on technical feasibility of Jatropha curcas oil based bio dielectric fluid for sustainable electric discharge machining (EDM). *Journal of Manufacturing Processes.* 22: 151–160.

[14] Leao FN, Pashby IR (2004) A review on the use of environmentally-friendly dielectric fluids in EDM. *Journal of Materials Processing Technology.* 149: 341–346.

[15] Zhu G, Bai J, Guo Y, Cao Y, Huang Y (2014) A study of the effects of working fluids on micro-hole arrays in micro-electrical discharge machining. *Proceedings of the Institution of Mechanical Engineers, Part B: Journal of Engineering Manufacture.* 228(11): 1381–1392.

[16] Chen SL, Yan BH, Huang FY (1999) Influence of kerosene and distilled water as dielectrics on the electric discharge machining characteristics of Ti-6A1-4V. *Journal of Materials Processing Technology.* 87: 107–111.

[17] Jeswani ML (1981) Effect of the addition of graphite powder to kerosene used as the dielectricfluid in electrical discharge machining. *Wear.* 70: 133–139.

[18] Jahan MP, Rahman M, Wong YS (2010) Modelling and experimental investigation on the effect of nano powder mixed dielectric in micro-electro discharge machining of tungsten carbide. *Proceedings of the Institution of Mechanical Engineers, Part B: Journal of Engineering Manufacture.* 224: 1725–1739.

[19] Ndaliman MB, Khan AA, Ali MY (2013) Influence of dielectric fluids on surface properties of electrical discharge machined titanium alloy. *Proceedings of the Institution of Mechanical Engineers, Part B: Journal of Engineering Manufacture.* 227(9): 1310–1316.

Chapter 3

Wire-fed arc-based additive manufacturing techniques and their recent advances

Bunty Tomar and Shiva S.
Indian Institute of Technology Jammu

3.1 INTRODUCTION

Researchers have consistently worked to improve the current manufacturing processes in order to provide a better methodology for the best use of money, resources, and time (Liu et al., 2019). In this decade, additive manufacturing (AM) technologies have become a viable method of creating three-dimensional components and quick prototypes. To create a complete 3-D component, AM technology employs a computerized setup in order to deposit material layer by layer (Milewski, 2017). This distinguishing characteristic of AM has made it known as a dependable and creative engineering process that creates the end-use geometry straight from the Computer Added Design (CAD) data, excluding the need for expensive, extensive tooling such as moulds, punches, or casting dies. Compared to traditional manufacturing methods, AM makes it easier to produce complicated and bespoke shapes. Additionally, it substantially lowers manufacturing costs, particularly for lower-volume products, and a considerably shorter cycle time may be anticipated (Dutta & Froes, 2016; Herzog et al., 2016; Huang et al., 2013). Part consolidation greatly reduces the quantity of pieces that need to be generated in AM in order to construct a finished component. Due to these innovative features, fewer conventional production processes and assembly lines are required (Sames et al., 2016). Additionally, compared to traditional manufacturing techniques, using AM can result in significant raw material savings (Lopes et al., 2020; Oliveira et al., 2020), for example, casting and forging methods (Yilmaz & Ugla, 2016). The primary advantages of AM over traditional methods for producing three-dimensional engineering analogues, such as computer numerical control machining, may be considered to be an improvement in structure excellence and performance and a solution to the problem of complex design (DebRoy et al., 2018). Due to a variety of such factors, AM is today regarded as the primary driver of the 4th industrial revolution and a common fabrication method for creating high-performance end-use components (Gibson et al., 2010; Khodabakhshi et al., 2020). Presently, components made with AM are used in increasingly sophisticated engineering and industrial applications across numerous

DOI: 10.1201/9781003406488-3

industries, including those in the automotive, aerospace, and medical sectors (Khodabakhshi & Gerlich, 2018; Li et al., 2019; Ngo et al., 2018).

As per the ASTM F2792 standard, AM is a process for creating 3D components by layer-by-layer combining of materials, as opposed to traditional subtractive manufacturing (ASTM F2792-12a, 2012). As per the ASTM F2792 standard, AM techniques can be categorized into seven primary categories: direct energy deposition (DED), powder bed fusion (PBF), sheet lamination (SL), binder jetting (BJ), material extrusion (ME), Material Jettting (MJ), and vat-photopolymerization (VP) (ASTM F2792-12a, 2012). VP, MJ, and ME are primarily employed in the manufacturing of polymer-based components, whereas SL and BJ can be employed for the fabrication of all grades of materials, including ceramics, metals, polymers, and composites. All of these processes are used, particularly for certain grades of materials. DED and PBF were specifically developed to produce three-dimensional metal components using an AM process (Gibson et al., 2015).

Today, arc-based additive manufacturing technologies are promising in the field of manufacturing as they excel ahead of other AM processes in terms of deposition rate, capital, time, and ease of operation. Using wire as feedstock in arc-based AM is another advantage over powder-based AM processes, as handling powder is a challenging affair and the cost of powder material is significantly higher than wire material (Tomar et al., 2022). Researchers and industry professionals are highly attracted to WAAM, but the build accuracy and need for post-processing operations are constraints in its commercialization. Recently, several advances have been made in the WAAM technique to minimize post-processing and make it more industry-oriented. This chapter presents an overview of the WAAM process and its advantages over laser and powder-based AM processes. It summarizes the recent advances in this technology, including support-free WAAM, ultracold WAAM, hot forging WAAM, near-immersion active cooling (NIAC) WAAM, etc., and presents them with their experimental method and validation.

3.2 WIRE ARC ADDITIVE MANUFACTURING

The principle of arc welding paired with a wire feed process is used in wire arc additive manufacturing (WAAM) technology. The earliest mention of WAAM dates back to the mid-1920s, when Baker Ralph submitted a US patent application for the production of metallic decorations utilizing an electric arc as an energy source (heat) and metal wire as the material (Ralph, n.d.). In WAAM, metallic wire feedstock is fed for melting by arc heat energy, and molten metal is deposited on the specific substrate utilizing beads in a layer-by-layer manner using data from a three-dimensional CAD model. A robotic or gantry system is used to deposit these beads continuously, layer by layer, resulting in the development of the entire 3D metal component (Tomar & Shiva, 2022). The WAAM approach has the

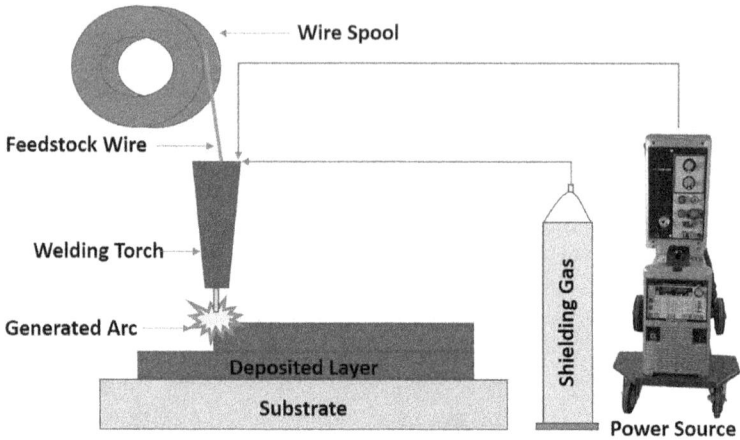

Figure 3.1 Graphic representation of a generalized wire-fed arc-based additive manufacturing system.

capacity to create application-based components with a near-net shape and surface roughness that is tolerably acceptable. It is necessary to machine end-use components using the proper machining techniques. A number of procedure selection factors, along with computer-aided process planning, must be taken into consideration in order to determine WAAM's capacity to manufacture almost flawless components made of certain materials. The parameters for CAD modelling, selection and generation of tool paths, selecting welding methods, optimizing welding technique parameters, selecting shielding gas composition, and selecting wire are the main selection factors for processes (Tomar et al., 2022).

A generalized WAAM system contains a power source, shielding gas cylinder, wire feeder, welding torch, and a substrate plate over which the deposition is carried out. A schematic representation of the WAAM process structure is presented in Figure 3.1.

3.3 CLASSIFICATION OF WAAM

The WAAM method may be roughly divided into three groups based on the kind of employed heat energy source: gas metal arc welding (GMAW) (Pattanayak & Sahoo, 2021), gas tungsten arc welding (GTAW) (Dickens et al., 1992), and plasma arc welding (PAW) (Spencer et al., 1998). Based on the user application, a particular welding method is chosen for a certain WAAM application. The deposition rate, amount of time required, and processing parameters chosen for a particular material are all directly impacted by the choice of the WAAM process. Pan et al. (2018) present these categories and variations in depth.

3.3.1 Gas metal arc welding (GMAW-WAAM)

In this process, energy is created when an electric arc is formed between the substrate and feed electrode (which has a positive terminal) and the negative of the power source. In GMAW-based WAAM, the manner of metal transfer is a crucial component since it directly affects the mechanical characteristics, microstructure, and surface quality (Tomar, 2020). A regulated mode of metal transfer and better control over deposit geometry and microstructure are being demonstrated via cold metal transfer (CMT)-based GMAW-WAAM (Tomar, 2023). The electrode in GMAW is protected by a shielding gas that is supplied outside (Nagamatsu et al., 2020; Pan et al., 2018). To investigate the impact of on-line milling on the precision of feedstock deposition, a milling setup integrated GMAW system, sometimes referred to as hybrid additive or subtractive manufacturing, is also created (Zhang et al., 2019). For the production of bimetallic and functionally graded materials, a tandem GMAW-WAAM based on a twin-wire feed mechanism offers a comparably greater rate of deposition and a simpler compositional gradient (Shi et al., 2019). It has been demonstrated that GMAW-based WAAM offers superior capabilities in terms of greater rate of deposition, better material usage, reduced production cost, and environmental friendliness in comparison with electron and laser beam-based, GTAW-based, and PAW-based AM techniques (Wang et al., 2017).

3.3.2 Gas tungsten arc welding (GTAW-WAAM)

In this process, an electric arc will form between the substrate and a non-consumable tungsten alloy electrode (with the negative terminal of the source). Through the use of a separate wire feeding mechanism, the feedstock material in wire form is made to enter the melt pool created by this arc via its leading-edge melting and depositing on the substrate (Ding et al., 2015). Less energy efficiency in the GTAW technique is caused by the anode's heat output, which is solely utilized for the formation of molten pools on the substrate's surface. The welding current is dependent on the welding electrode size; a larger electrode allows for a higher current, and vice versa (Sathishkumar & Manikandan, 2019). GTAW-based processes have the lowest deposition rates in comparison to GMAW- and PAW-based WAAM techniques.

3.3.3 Plasma arc welding (PAW-WAAM)

In this technique, a tungsten non-consumable electrode and a water-cooled nozzle are used to create the arc. When an inert gas (most often Ar) is passed into the arcing zone, it ionizes and produces a jet of plasma. Through a small nozzle aperture, the produced plasma jet is directed to the clamped substrate plate, causing a melt pool to form on top of the substrate. Through a different wire feeder mechanism, a feedstock material of the wire form is fed into this molten pool, where the feedstock melts and is deposited at the substrate

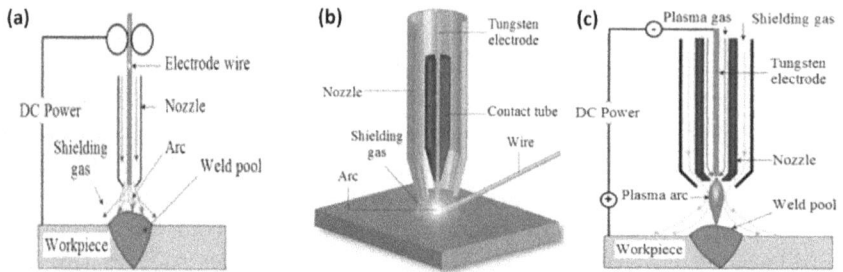

Figure 3.2 Schematic representation of (a) GMAW-, (b) GTAW-, and (c) PAW-based techniques (Jafari et al., 2021) (open access).

surface (Artaza et al., 2020). PAW's deposition rates are of intermediate value since they are greater than GTAW's but lower than GMAW's deposition rates. The maximum energy density of the electrical arc is provided by PAW, enabling faster travel times and minimal distortion in the weld beads (Bölmsjo, 2006). Although PAW-WAAM is capable of fabricating welds with the highest geometrical and structural excellence, extensive capital is required to set up its process setup. Figure 3.2 presents a schematic representation of the basic principles and workings of GMAW-, GTAW-, and PAW-based processes.

3.4 ADVANTAGES OF WAAM OVER LASER AND POWDER-BASED ADDITIVE MANUFACTURING (AM)

The utilization of an electric arc in WAAM creates a fusion source that is more efficient than previous AM methods (Baumers et al. (n.d.); Jackson et al., 2016). From the perspective of energy consumption, using more efficient fusion sources in WAAM is advantageous, especially for reflective metals with low coupling efficiency for laser beams, for example aluminium (Leirmo & Baturynska, 2020), copper (Gu et al., 2012), and magnesium (Guo et al., 2016). These arc-based AM methods, however, give the deposit a relatively larger energy input. In addition to that, employing wire as feedstock in WAAM reduces the necessity of the powder recycling process (Moghimian et al., 2021; Tang et al., 2015), improving operating conditions for operators. A comparison of arc-based AM techniques with laser beam- and electron beam-based AM techniques is presented in Table 3.1. For many engineering materials, using wire results in a significant per-kilogram price reduction compared to using powder, such as Ti64, IN625, and SS316L wires, which have a price of Rs. 10,500, 4,400, and 1,100 per kg in comparison to a powder price of Rs. 25,000, 7,000, and 3,500 per kg, respectively (Tomar et al., 2022).

Table 3.1 Comparison of arc AM with laser and electron beam AM methods (Tomar et al., 2022)

Parameter	Laser based	Electron based	Arc based
Heat source	Electromagnetic wave	Kinetic energy	Electric arc
Energy density	10^6 W/mm^2	10^8 W/mm^2	10^6–10^8 W/mm^2
Energy efficiency	Least	Intermediate	Excellent
Deposition rate	150–200 g/h	600–800 g/h	2–4 kg/h
Vacuum required	No	Yes	No

Furthermore, the parts fabricated with wire-fed arc-based additive manufacturing techniques have a relatively higher density and don't have a size limitation; thus, large-sized component fabrication is possible via this route. Using wire as the feedstock provides an additional advantage regarding material quality, as there is less chance of contamination in the wire as compared to powder (Williams et al., 2016). Numerous improvements, adjustments, and studies have been made to WAAM since its creation in 1925, but it hasn't yet reached the full potential of industrial commercialization globally.

3.5 RECENT ADVANCES IN WAAM

Components fabricated using the WAAM technique need the right post-processing procedures to enhance the component's characteristics and get rid of major problems such as surface abrasion, porosity, and residual tensions. In addition, a number of improvements are made to the way WAAM operates, its tooling, and its workings in order to address various process difficulties in the production of WAAM-based materials. In this portion of the study, an overview of these post-processing methods and current developments in WAAM is provided.

3.5.1 Near-immersion active cooling

The major difficulty in WAAM-based applications is thermal control. The continuous cyclic heating and cooling process, along with the interpass temperature encountered during deposition, have a significant impact on the microstructure and mechanical characteristics of a WAAM-fabricated structure as they are formed (Vahedi Nemani et al., 2021). In the WAAM approach, NIAC was recently created and tested. NIAC aims to improve the process of thermal management in the wire and arc AM to reduce heat buildup brought on by excessive heat input. The NIAC concept was presented by da Silva et al. (2021) based on the active cooling of the previously formed structure by immersing the deposit in a cooling liquid in a prescribed and controlled fashion.

By reducing the buildup of heat during the WAAM manufacturing of the AWS ER5356 alloy, da Silva et al. (2020) confirmed the use of the NIAC for their research. The experimental setup for the NIAC idea is depicted in Figure 3.3b. Figure 3.3 depicts three distinct methods of doing the deposition that were used to demonstrate the viability of the NIAC idea (b). In comparison to the other two techniques, the outcomes indicated that the walls that NIAC deposited had the best geometric quality and the lowest percentage of voids and porosity. The NIAC process has been shown to have better mechanical properties with less anisotropy. Scotti et al. (2020) published similar results for thermal management in CMT WAAM manufactured structures employing the discussed NIAC approach.

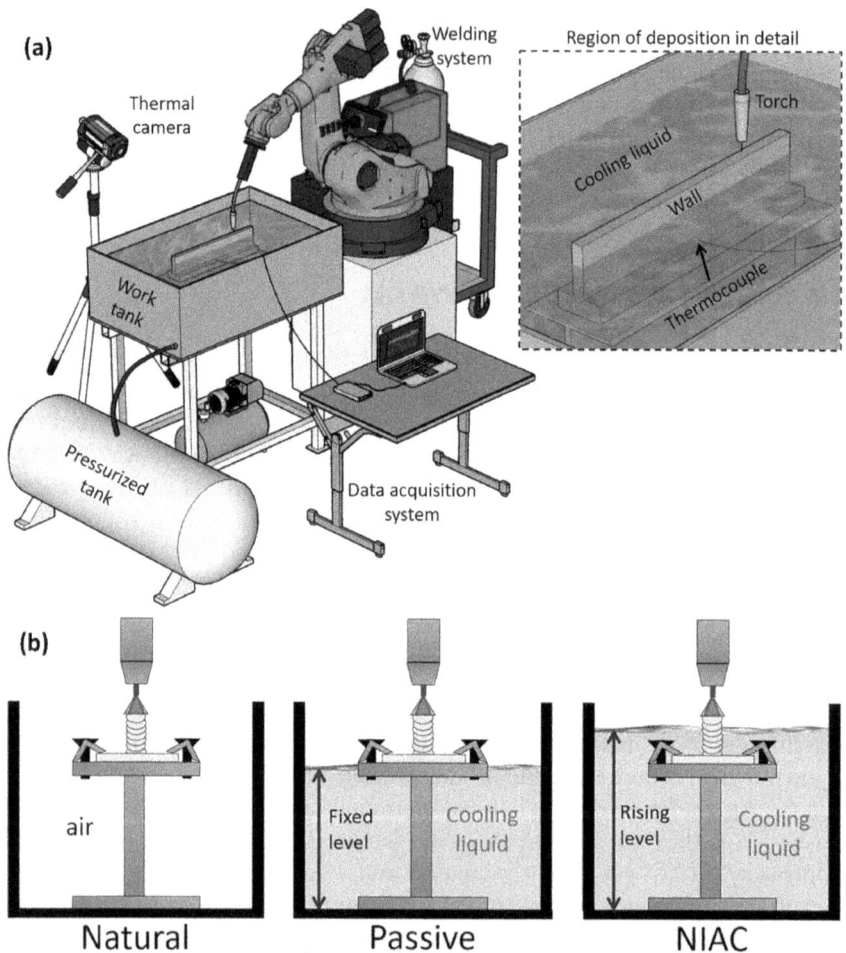

Figure 3.3 Schematic representation for (a) laboratory rig arrangement for NIAC and (b) cooling approaches employed in this experiment (da Silva et al., 2020). Reprinted with permission of Springer Nature.

3.5.2 Interpass cooling setup

Components made using WAAM have a complicated thermal profile that involves several re-heating and re-cooling cycles, which leads to complex grain structures and subpar mechanical qualities. Interpass cooling methods have recently been developed using forced interpass cooling employing compressed CO_2 gas to achieve improved microstructure and improved mechanical characteristics. The in situ interlayer temperatures and the thermal cycles may be changed into a suitable domain for attaining the necessary microstructure and mechanical characteristics. Fast cooling is accomplished by utilizing a gas-supplying, movable nozzle. Figure 3.4 shows a schematic representation of the integrated WAAM with a gas cooling configuration.

An initial investigation by Wu et al. (2018) for wire arc additively manufacturing assisted production of Ti-64 components employing forced interpass cooling by compressed CO_2 produced promising results. It was discovered that the interpass cooling technique results in a better microstructure, reduced oxidation, increased hardness, and greater strength. Le et al. (2021) did experimentation on the use of active cooling involving compressed dry air in the WAAM production of SS308L structures to verify and validate the active cooling technique. Compared to components made without the use of active cooling, favourable outcomes were attained with the least surface roughness, highest average hardness, and improved mechanical strength. Although this technique yielded good results, there is a need for more research in this area.

3.5.3 Interpass cold rolling setup

The complex heating and cooling cycles in the WAAM method produce anisotropic microstructure and mechanical characteristics as a result of the temperature gradient present within the deposit and complicated warming and re-cooling phenomena. In order to greatly minimize the anisotropy in the

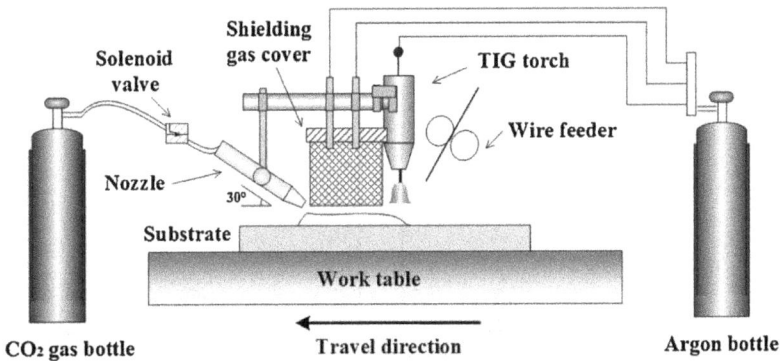

Figure 3.4 Schematic presentation of interpass cooling integrated wire arc additive manufacturing setup (Wu et al., 2018). Reprinted with permission of Elsevier.

microstructure through the process of permanent deformation of the deposit, interpass cold rolling of the deposited structure has developed. The amount of residual stress and deformation may be significantly reduced by interpass rolling the weld bead (Colegrove et al., 2017). In 2013, Colegrove et al. (2013) utilized this idea to create the WAAM system integrated with a cold rolling setup. Figure 3.5 illustrates how a hydraulically loaded slotted roller is designed to travel over the deposit structure at the same constant speed as the GMAW welding flame. The finished component required practically no machining since the roller's use was successful in decreasing the deposited structure's distortion and surface roughness. In succeeding layers of deposition, the warming of the rolled layer caused advantageous grain refining.

The idea of interpass rolling in wire arc additively for the manufacturing of Ti6Al4V alloys was proven by Hu et al. (2021). The results were significantly better in comparison to as-built WAAM specimens, with UTS and YS of 876 and 789 MPa, respectively, and a strain-to-fracture percentage of 11%. Figure 3.6 illustrates how in situ rolling of the deposit reduced voids and increased dislocation density, leading to a gain in strength and ductility. Zhang et al. (2021) observed similar outcomes in WAAM-based

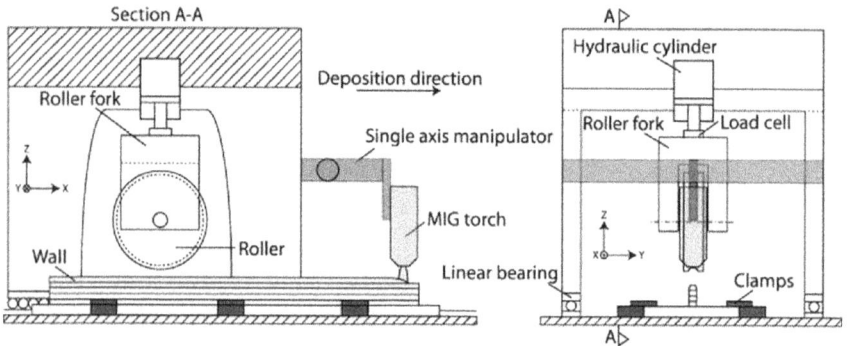

Figure 3.5 Diagrammatic representation of interpass cold rolling-integrated WAAM setup (Colegrove et al., 2013). Reprinted with permission of Elsevier.

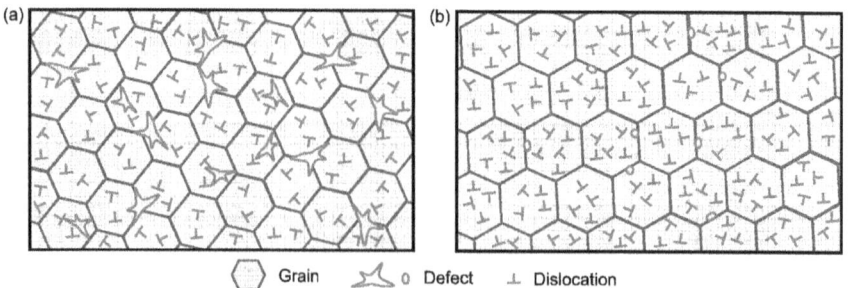

Figure 3.6 Schematic representation of microstructure formed in manufacturing by (a) WAAM setup and (b) cold rolling-integrated WAAM setup (Hu et al., 2021). Reprinted with permission of Elsevier.

manufacturing of IN718 material. This method is currently only used to create straightforward, thin walls. Given that rolling techniques have geometric restrictions, there is still a lot of research to be carried out before in situ rolling is used in the additive manufacturing of ready-to-use components.

3.5.4 Hot forging WAAM (HF-WAAM)

At the University of Caparica, Duarte et al. (2020) recently invented the innovative idea for "Hot forging wire arc additive manufacturing" (HF-WAAM). This method's main goal is to make use of the materials' high-temperature viscoplastic deformation tendency to improve strength, lower residual stress, and homogenize grain structure. A vibrating actuator that is either mechanically or electromagnetically powered strikes a hammer inside the gas nozzle as part of HF-WAAM. This operation's deformation is similar to the hot forging technique. Figure 3.7a and b shows a schematic illustration of the forging and its mechanism, respectively.

In the first investigations, manufacturing of AISI316L stainless steel produced positive results. When compared to as-fabricated WAAM samples, enhanced testing findings are exhibited in Figure 3.8 due to a more refined

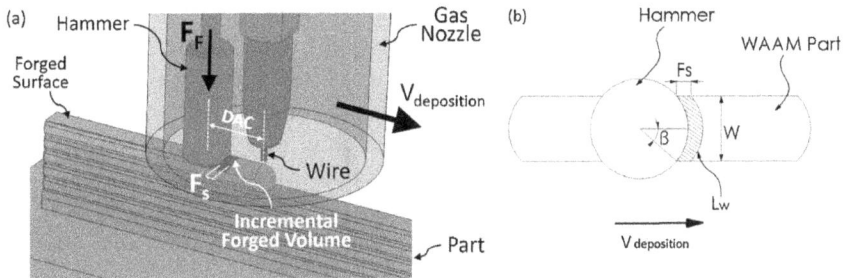

Figure 3.7 Schematical presentation of the forged region in Hot Forging WAAM: (a) isometric view (3D) and (b) top view (2D) (Li et al., 2019) (open access).

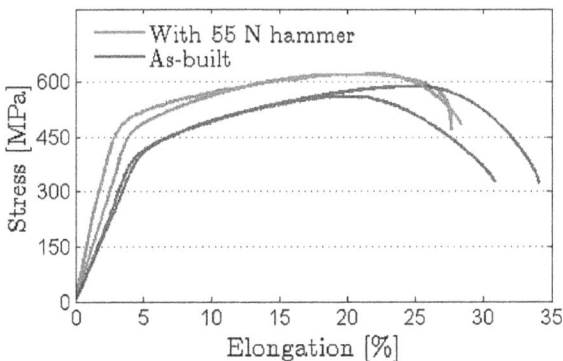

Figure 3.8 Stress vs. elongation graph for tensile results of as-fabricated WAAM deposit and 55 N hammered WAAM deposit (Li et al., 2019) (open access).

grain structure and void closure through hot forging. More study must be done before HF-WAAM may be commercialized, despite having the potential to be employed in industrial applications.

3.5.5 Ultracold WAAM (UC-WAAM)

The handling of excessive heat input, which results in higher levels of residual stresses and distortions, is the main issue in the WAAM process, according to a comprehensive analysis of the literature. CMT-based WAAM and HF-WAAM are two methods to decrease the heat accumulation in WAAM without any compromise to the productivity of the operation (Li et al., 2019). Recently, Rodrigues et al. (2021) created a novel wire and arc-based AM technique known by the name "Ultracold Wire Arc Additive Manufacturing." In the UC-WAAM, feedstock metal wire and an inert tungsten electrode were what caused the electric arc to form. This modification was intended to raise the cooling rate, cut off the power supply (current) from reaching the substrate plate, and keep the electricity away from the melt pool in order to lower the process temperature. Figure 3.9 compares the schematic representation of the ultracold WAAM with the GMAW-WAAM.

A feasibility study on the manufacturing of HSLA steel using UC-WAAM compared to the GMAW-based WAAM system was conducted to validate this technology. Better arc stability, along with aesthetic wall appearance and avoidance of spatter development, showed encouraging outcomes. This method helped in accelerating the cooling rate by 11°C/s and had a substantial impact on the formation of the microstructure. The researchers recommended using UC-WAAM to fabricate self-supporting overhanging components, although more extensive study has to be done in this area.

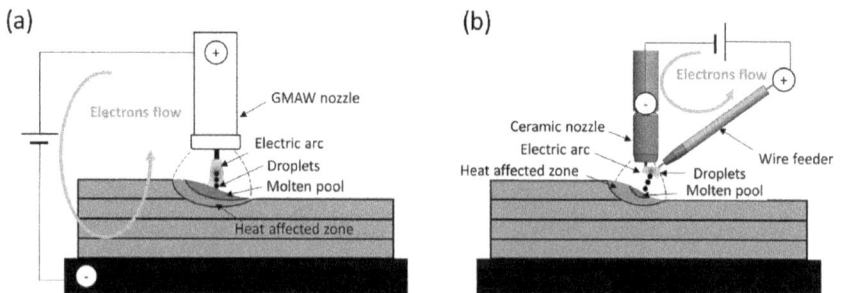

Figure 3.9 Schematical presentation of the arc generation phenomenon in (a) GMAW-WAAM and (b) ultracold WAAM (Rodrigues et al., 2021). Reprinted with permission of Elsevier.

3.5.6 Support-free WAAM

In metal AM, removing the support structure is a difficult process since it impairs the component's visual appeal and calls for more machining and longer deposition times. The development of support-free AM techniques has attracted a lot of researchers' attention. Vanek et al. (2014) suggested a geometry-based solution that involves rotating the model so that the support structures take up the least amount of space. To split the design into roughly pyramidal forms and have them printed independently to avoid supports, Hu et al. (2014) introduced a segmentation model. Because of the existence of empty gaps between the layers, these models' layer segmentation led to cumulative fitness errors.

Recently, Liu et al. (2021) suggested a technique of surface segmentation and surface separation based on multi-degree of freedom depositing channels to facilitate support-free wire and arc-based AM. In this approach, the CAD model was later segmented into different zones named surface and internal solid (IS), surface's overhanging area (OA), and non-overhanging area (NOA). While OA was placed using the multi-DOF deposition technique, IS and NOA were deposited using the conventional vertical WAAM deposition technique. In order to validate two example models, this new support-free WAAM approach was deposited. Although both models yielded stable deposition, several cumulative defects in layer thickness and wall inclination were found. This strategy needs more investigation in the areas of deposition uniformity and thermal deformation control.

3.6 CONCLUSIONS

Based on the excellent potential and capabilities of wire-fed arc-based AM techniques, it has captured the attention of industries and researchers to use it as a mainstream AM technique. In this chapter, an overview of wire-fed arc-based AM processes with emphasis on their recent technical developments is presented. This chapter starts with the introduction of AM techniques and highlights the need for arc-based AM technologies by addressing the advantages of arc-based AM techniques over contemporary AM methods. WAAM was detailed with its different variants, including GMAW-, GTAW-, and PAW-based WAAM techniques. Controlling the higher heat input in WAAM is a major challenge in arc-based AM techniques, so various recent advances in this field, including NIAC, interpass cooling setup, interpass cold rolling setup, hot forging WAAM, ultracold WAAM, and support-free WAAM, were discussed in the last portion of this chapter.

REFERENCES

Artaza, T., Suárez, A., Veiga, F., Braceras, I., Tabernero, I., Larrañaga, O., & Lamikiz, A. (2020). Wire arc additive manufacturing Ti6Al4V aeronautical parts using plasma arc welding: analysis of heat-treatment processes in different atmospheres. *Journal of Materials Research and Technology*, 9(6), 15454–15466. https://doi.org/10.1016/j.jmrt.2020.11.012.

ASTM F2792-12a. (2012). *Standard Terminology for Additive Manufacturing Technologies*, ASTM International, West Conshohocken, PA. ASTM International.

Tomar, B. (2020). Fatigue analyses of GMAW welds of thermo-mechanically processed 700MC ultra high strength steel. *IJRESM*, 3(3), 220–225.

Baumers, M., Tuck, C., Hague, R., Ashcroft, I., & Wildman, R. (n.d). A comparative study of metallic additive manufacturing power consumption, 21st Annual International Solid Freeform Fabrication Symposium - An Additive Manufacturing Conference. SFF, University of Texas. https://dx.doi.org/10.26153/tsw/15198.

Bölmsjo, J. P. A. L. G. (2006). *Welding Robots*. Springer-Verlag, London. https://doi.org/10.1007/1-84628-191-1.

Colegrove, P. A., Coules, H. E., Fairman, J., Martina, F., Kashoob, T., Mamash, H., & Cozzolino, L. D. (2013). Microstructure and residual stress improvement in wire and arc additively manufactured parts through high-pressure rolling. *Journal of Materials Processing Technology*, 213(10), 1782–1791. https://doi.org/10.1016/j.jmatprotec.2013.04.012.

Colegrove, P. A., Donoghue, J., Martina, F., Gu, J., Prangnell, P., & Hönnige, J. (2017). Application of bulk deformation methods for microstructural and material property improvement and residual stress and distortion control in additively manufactured components. *Scripta Materialia*, 135, 111–118. https://doi.org/10.1016/j.scriptamat.2016.10.031.

daSilva, L. J., Ferraresi, H. N., Araújo, D. B., Reis, R. P., & Scotti, A. (2021). Effect of thermal management approaches on geometry and productivity of thin-walled structures of ER 5356 built by wire+arc additive manufacturing. Coatings, 11(9), 1141. https://doi.org/10.3390/coatings11091141.

daSilva, L. J., Souza, D. M., de Araújo, D. B., Reis, R. P., & Scotti, A. (2020). Concept and validation of an active cooling technique to mitigate heat accumulation in WAAM. *The International Journal of Advanced Manufacturing Technology*, 107(5–6), 2513–2523. https://doi.org/10.1007/s00170-020-05201-4.

DebRoy, T., Wei, H. L., Zuback, J. S., Mukherjee, T., Elmer, J. W., Milewski, J. O., Beese, A. M., Wilson-Heid, A., De, A., & Zhang, W. (2018). Additive manufacturing of metallic components - process, structure and properties. *Progress in Materials Science*, 92, 112–224. https://doi.org/10.1016/j.pmatsci.2017.10.001.

Ding, D., Pan, Z., Cuiuri, D., & Li, H. (2015). Wire-feed additive manufacturing of metal components: technologies, developments and future interests. *International Journal of Advanced Manufacturing Technology*, 81(1–4), 465–481. https://doi.org/10.1007/s00170-015-7077-3.

Duarte, V. R., Rodrigues, T. A., Schell, N., Miranda, R. M., Oliveira, J. P., & Santos, T. G. (2020). Hot forging wire and arc additive manufacturing (HF-WAAM). *Additive Manufacturing*, 35, 101193. https://doi.org/10.1016/j.addma.2020.101193.

Dutta, B., & Froes, F. H. (2016). The Additive Manufacturing of Titanium Alloys. In B. Dutta & F. H. Froes (eds.) *Additive Manufacturing of Titanium Alloys* (pp. 1–10). Elsevier, Butterworth-Heinemann. https://doi.org/10.1016/B978-0-12-804782-8.00001-X.

Gibson, I., Rosen, D., & Stucker, B. (2015). *Additive Manufacturing Technologies.* Springer, New York. https://doi.org/10.1007/978-1-4939-2113-3.

Gibson, I., Rosen, D. W., & Stucker, B. (2010). *Additive Manufacturing Technologies.* Springer, US. https://doi.org/10.1007/978-1-4419-1120-9.

Gu, D. D., Meiners, W., Wissenbach, K., & Poprawe, R. (2012). Laser additive manufacturing of metallic components: materials, processes and mechanisms. *International Materials Reviews*, 57(3), 133–164. https://doi.org/10.1179/1743280411Y.0000000014.

Guo, J., Zhou, Y., Liu, C., Wu, Q., Chen, X., & Lu, J. (2016). Wire arc additive manufacturing of AZ31 magnesium alloy: grain refinement by adjusting pulse frequency. *Materials*, 9(10), 823. https://doi.org/10.3390/ma9100823.

Herzog, D., Seyda, V., Wycisk, E., & Emmelmann, C. (2016). Additive manufacturing of metals. *Acta Materialia*, 117, 371–392. https://doi.org/10.1016/j.actamat.2016.07.019.

Hu, R., Li, H., Zhang, H., & Cohen-Or, D. (2014). Approximate pyramidal shape decomposition. *ACM Transactions on Graphics*, 33(6), 1–12. https://doi.org/10.1145/2661229.2661244.

Hu, Y., Ao, N., Wu, S., Yu, Y., Zhang, H., Qian, W., Guo, G., Zhang, M., & Wang, G. (2021). Influence of in situ micro-rolling on the improved strength and ductility of hybrid additively manufactured metals. *Engineering Fracture Mechanics*, 253, 107868. https://doi.org/10.1016/j.engfracmech.2021.107868.

Huang, S. H., Liu, P., Mokasdar, A., & Hou, L. (2013). Additive manufacturing and its societal impact: a literature review. *The International Journal of Advanced Manufacturing Technology*, 67(5–8), 1191–1203. https://doi.org/10.1007/s00170-012-4558-5.

Jackson, M. A., Van Asten, A., Morrow, J. D., Min, S., & Pfefferkorn, F. E. (2016). A comparison of energy consumption in wire-based and powder-based additive-subtractive manufacturing. *Procedia Manufacturing*, 5, 989–1005. https://doi.org/10.1016/j.promfg.2016.08.087.

Jafari, D., Vaneker, T. H. J., & Gibson, I. (2021). Wire and arc additive manufacturing: opportunities and challenges to control the quality and accuracy of manufactured parts. *Materials & Design*, 202, 109471. https://doi.org/10.1016/j.matdes.2021.109471.

Khodabakhshi, F., Farshidianfar, M. H., Gerlich, A. P., Nosko, M., Trembošová, V., & Khajepour, A. (2020). Effects of laser additive manufacturing on microstructure and crystallographic texture of austenitic and martensitic stainless steels. *Additive Manufacturing*, 31, 100915. https://doi.org/10.1016/j.addma.2019.100915.

Khodabakhshi, F., & Gerlich, A. P. (2018). Potentials and strategies of solid-state additive friction-stir manufacturing technology: a critical review. *Journal of Manufacturing Processes*, 36, 77–92. https://doi.org/10.1016/j.jmapro.2018.09.030.

Le, V. T., Mai, D. S., & Paris, H. (2021). Influences of the compressed dry air-based active cooling on external and internal qualities of wire-arc additive manufactured thin-walled SS308L components. *Journal of Manufacturing Processes*, 62, 18–27. https://doi.org/10.1016/j.jmapro.2020.11.046.

Leirmo, J. L., & Baturynska, I. (2020). Challenges and proposed solutions for aluminium in laser powder bed fusion. *Procedia CIRP*, 93, 114–119. https://doi.org/10.1016/j.procir.2020.03.090.

Li, N., Huang, S., Zhang, G., Qin, R., Liu, W., Xiong, H., Shi, G., & Blackburn, J. (2019). Progress in additive manufacturing on new materials: a review. *Journal of Materials Science & Technology*, 35(2), 242–269. https://doi.org/10.1016/j.jmst.2018.09.002.

Li, Z., Liu, C., Xu, T., Ji, L., Wang, D., Lu, J., Ma, S., & Fan, H. (2019). Reducing arc heat input and obtaining equiaxed grains by hot-wire method during arc additive manufacturing titanium alloy. *Materials Science and Engineering A*, 742, 287–294. https://doi.org/10.1016/j.msea.2018.11.022.

Liu, B., Shen, H., Zhou, Z., Jin, J., & Fu, J. (2021). Research on support-free WAAM based on surface/interior separation and surface segmentation. *Journal of Materials Processing Technology*, 297, 117240. https://doi.org/10.1016/j.jmatprotec.2021.117240.

Liu, W., Jia, C., Guo, M., Gao, J., & Wu, C. (2019). Compulsively constricted WAAM with arc plasma and droplets ejected from a narrow space. *Additive Manufacturing*, 27, 109–117. https://doi.org/10.1016/j.addma.2019.03.003.

Lopes, J. G., Machado, C. M., Duarte, V. R., Rodrigues, T. A., Santos, T. G., & Oliveira, J. P. (2020). Effect of milling parameters on HSLA steel parts produced by wire and arc additive manufacturing (WAAM). *Journal of Manufacturing Processes*, 59, 739–749. https://doi.org/10.1016/j.jmapro.2020.10.007.

Milewski, J. O. (2017). *Additive Manufacturing of Metals* (Vol. 258). Springer International Publishing, Cham. https://doi.org/10.1007/978-3-319-58205-4.

Moghimian, P., Poirié, T., Habibnejad-Korayem, M., Zavala, J. A., Kroeger, J., Marion, F., & Larouche, F. (2021). Metal powders in additive manufacturing: a review on reusability and recyclability of common titanium, nickel and aluminum alloys. *Additive Manufacturing*, 43, 102017. https://doi.org/10.1016/j.addma.2021.102017.

Nagamatsu, H., Sasahara, H., Mitsutake, Y., & Hamamoto, T. (2020). Development of a cooperative system for wire and arc additive manufacturing and machining. *Additive Manufacturing*, 31, 100896. https://doi.org/10.1016/j.addma.2019.100896.

Ngo, T. D., Kashani, A., Imbalzano, G., Nguyen, K. T. Q., & Hui, D. (2018). Additive manufacturing (3D printing): a review of materials, methods, applications and challenges. *Composites Part B: Engineering*, 143, 172–196. https://doi.org/10.1016/j.compositesb.2018.02.012.

Oliveira, J. P., LaLonde, A. D., & Ma, J. (2020). Processing parameters in laser powder bed fusion metal additive manufacturing. *Materials & Design*, 193, 108762. https://doi.org/10.1016/j.matdes.2020.108762.

Dickens, P., Pridham, M., Cobb, R., Gibson, I., & Dixon, G. (1992). Rapid prototyping using 3-D welding, In DTIC Document. https://hdl.handle.net/2152/64409.

Pan, Z., Ding, D., Wu, B., Cuiuri, D., Li, H., & Norrish, J. (2018). *Arc Welding Processes for Additive Manufacturing: A Review*, Shanben Chen, Yuming Zhang, Zhili Feng (eds.), *Transactions on Intelligent Welding Manufacturing* (pp. 3–24). Springer, Singapore. https://doi.org/10.1007/978-981-10-5355-9_1.

Pattanayak, S., & Sahoo, S. K. (2021). Gas metal arc welding based additive manufacturing-a review. *CIRP Journal of Manufacturing Science and Technology*, 33, 398–442. https://doi.org/10.1016/j.cirpj.2021.04.010.

Ralph, B. (n.d.). Method of making decorative articles. United States, US1533300A. Filed Nov. 12, 1920, Granted April 14, 1925.

Rodrigues, T. A., Duarte, V. R., Miranda, R. M., Santos, T. G., & Oliveira, J. P. (2021). Ultracold-wire and arc additive manufacturing (UC-WAAM). *Journal of Materials Processing Technology*, 296, 117196. https://doi.org/10.1016/j.jmatprotec.2021.117196.

Sames, W. J., List, F. A., Pannala, S., Dehoff, R. R., & Babu, S. S. (2016). The metallurgy and processing science of metal additive manufacturing. *International Materials Reviews*, 61(5), 315–360. https://doi.org/10.1080/09506608.2015.1116649.

Sathishkumar, M., & Manikandan, M. (2019). Preclusion of carbide precipitates in the Hastelloy X weldment using the current pulsing technique. *Journal of Manufacturing Processes*, 45, 9–21. https://doi.org/10.1016/j.jmapro.2019.06.027.

Scotti, F. M., Teixeira, F. R., Silva, L. J. da, de Araújo, D. B., Reis, R. P., & Scotti, A. (2020). Thermal management in WAAM through the CMT Advanced process and an active cooling technique. *Journal of Manufacturing Processes*, 57, 23–35. https://doi.org/10.1016/j.jmapro.2020.06.007.

Shi, J., Li, F., Chen, S., Zhao, Y., & Tian, H. (2019). Effect of in-process active cooling on forming quality and efficiency of tandem GMAW-based additive manufacturing. *The International Journal of Advanced Manufacturing Technology*, 101(5–8), 1349–1356. https://doi.org/10.1007/s00170-018-2927-4.

Spencer, J. D., Dickens, P. M., & Wykes, C. M. (1998). Rapid prototyping of metal parts by three-dimensional welding. *Proceedings of the Institution of Mechanical Engineers, Part B: Journal of Engineering Manufacture*, 212(3), 175–182. https://doi.org/10.1243/0954405981515590.

Tang, H. P., Qian, M., Liu, N., Zhang, X. Z., Yang, G. Y., & Wang, J. (2015). Effect of powder reuse times on additive manufacturing of Ti-6Al-4V by selective electron beam melting. *JOM*, 67(3), 555–563. https://doi.org/10.1007/s11837-015-1300-4.

Tomar, B. (2023). Cold metal transfer-based wire arc additive manufacturing. *Journal of the Brazilian Society of Mechanical Sciences and Engineering*, 2, 159–173. https://doi.org/10.1007/s40430-023-04084-2.

Tomar, B., & Shiva, S. (2022). Microstructure evolution in steel/copper graded deposition prepared using wire arc additive manufacturing. *Materials Letters*, 328, 133217. https://doi.org/10.1016/j.matlet.2022.133217.

Tomar, B., Shiva, S., & Nath, T. (2022). A review on wire arc additive manufacturing: processing parameters, defects, quality improvement and recent advances. *Materials Today Communications*, 31, 103739. https://doi.org/10.1016/j.mtcomm.2022.103739.

Vahedi Nemani, A., Ghaffari, M., Salahi, S., Lunde, J., & Nasiri, A. (2021). Effect of interpass temperature on the formation of retained austenite in a wire arc additive manufactured ER420 martensitic stainless steel. *Materials Chemistry and Physics*, 266, 124555. https://doi.org/10.1016/j.matchemphys.2021.124555.

Vanek, J., Galicia, J. A. G., & Benes, B. (2014). Clever support: efficient support structure generation for digital fabrication. *Eurographics Symposium on Geometry Processing*, 33(5), 117–125. https://doi.org/10.1111/cgf.12437.

Wang, Y., Chen, X., & Konovalov, S. V. (2017). Additive manufacturing based on welding arc: a low-cost method. *Journal of Surface Investigation: X-Ray, Synchrotron and Neutron Techniques*, 11(6), 1317–1328. https://doi.org/10.1134/S1027451017060210.

Williams, S. W., Martina, F., Addison, A. C., Ding, J., Pardal, G., & Colegrove, P. (2016). Wire+arc additive manufacturing. *Materials Science and Technology*, 32(7), 641–647. https://doi.org/10.1179/1743284715Y.0000000073.

Wu, B., Pan, Z., Ding, D., Cuiuri, D., Li, H., & Fei, Z. (2018). The effects of forced interpass cooling on the material properties of wire arc additively manufactured Ti6Al4V alloy. *Journal of Materials Processing Technology*, 258, 97–105. https://doi.org/10.1016/j.jmatprotec.2018.03.024.

Yilmaz, O., & Ugla, A. A. (2016). Shaped metal deposition technique in additive manufacturing: a review. *Proceedings of the Institution of Mechanical Engineers, Part B: Journal of Engineering Manufacture*, 230(10), 1781–1798. https://doi.org/10.1177/0954405416640181.

Zhang, S., Zhang, Y., Gao, M., Wang, F., Li, Q., & Zeng, X. (2019). Effects of milling thickness on wire deposition accuracy of hybrid additive/subtractive manufacturing. *Science and Technology of Welding and Joining*, 24(5), 375–381. https://doi.org/10.1080/13621718.2019.1595925.

Zhang, T., Li, H., Gong, H., Wu, Y., Ahmad, A. S., & Chen, X. (2021). Effect of rolling force on tensile properties of additively manufactured Inconel 718 at ambient and elevated temperatures. *Journal of Alloys and Compounds*, 884, 161050. https://doi.org/10.1016/j.jallcom.2021.161050.

Chapter 4

Joining of metal to polymers by hybrid additive manufacturing methods

Tharmaraj R
SRM Institute of Science and Technology

Rajesh Jesudoss Hynes N
Opole University of Technology Poland
Mepco Schlenk Engineering College

Shenbaga Velu P
Vellore Institute of Technology

4.1 INTRODUCTION

4.1.1 Need hybrid metal-plastic joints

Hybrid metal-plastic couplings are becoming more common in the electronics, automotive, biomedical, and aerospace sectors because of their ability to generate lightweight yet durable constructions. For instance, the auto industry's shift to environmentally friendly lightweight materials has resulted in a rise in the usage of plastics [1]. Plastics are made up of great particles known as polymers. Polymers have good features in thermal expansion, fatigue strength, fracture strength, and specific strength. Moreover, polymers are supple and lightweight materials. Due to their properties, polymers are sought-after materials, in the aerospace and automotive fields [2]. The capacity to connect metals and polymers is required for that. An item made from such an assemblage would have the strength and flexibility of metals in addition to the lightweight characteristics of polymers.

4.1.2 Metal-to-polymer joining methods

Metals and polymers are mixed to form hybrid structures that are both strong and light, as was already mentioned. Today, a variety of methods are used to attach polymers to metals, including adhesive bonding, mechanical fastening, and direct joining procedures. First off, a polymeric adhesive ingredient is used in the traditional joining method known as adhesive bonding to create a junction between the connecting elements. The strength of the

DOI: 10.1201/9781003406488-4

junction is mostly determined by the inter-particle forces among the linking components and the adhesive substance. Prior to bonding, surface pre-treatment of the components is crucial to boosting the strength and longevity of the junction. Adhesive bonding is utilized in an extensive variety of practices because it may offer lightweight joints and reliable stress circulation when exposed to weight [3]. The strength of adhesive joints was the subject of numerous investigations [4–8]. Yet, there are a number of downsides to adhesive bonds. One such restriction is their reactivity to ecological issues, particularly temperature and humidity. Moreover, rather than degrading over time, formed bonded joints frequently collapse right away [3].

Mechanical fastening is another easy joining method. Even though there are a variety of methods and materials utilized in mechanical fastening, riveting is the main subject of research because it has been shown to produce solid joints when combining metal and polymers [9–12]. Since it is exposed to the most deformation, the metallic component is frequently positioned at the bottom of the assembly, with the polymer sitting on top of it. The junction is then created by drilling a rivet through the plastic and inside the metal. However, the foremost disadvantages of such an approach are the heavier shape as a result of the outside rivet and the higher stress attention close to the rivet pits [3].

Alternative joining techniques have been studied and developed more and more over the past few decades as a result of the rising demand for metal-polymer mixture assemblies in manufacturing and the shortcomings of normal connection techniques. In order to dissolve the oxide layer and make the connection, ultrasonic welding (USW) uses high-frequency, low-amplitude waves to be used on the weld portion while it is dense [13–15]. USW stands out for being reasonably priced and completing orders quickly. But, in the case of metal-polymer bonding, the quality of the materials manufactured through the USW is affected due to the large reactivity of the waves applied [15].

Another method for combining metal and polymers is laser joining, which entails exposing the connection to a laser beam and forcing the foams of the polymeric component to open out and circulate in the melted dense connection, creating the link between the metal and the plastics. Robust joints are created as a result of the chemical bond created during laser welding between the oxide layer and carbon particles, as well as the mechanical connections formed as an outcome of van der Waals forces. The use of shielding gas and complex procedures are the main limitations of laser welding [3,16].

Another method of combining metal and polymer is injection moulding, in which the polymer is injected into the mould via a nozzle. After the mould has been full, compression is utilized to push the softening and prevent contraction in solidification [17–20]. Such a technique stands out for its fast production of complex structures. However, two of the main limitations of injection moulding are the low weld quality and the requirement for additive pre-treatment [19].

Another type of metal-polymer joining is called friction-assisted joining, in which a device inserts the metal portion into the plastic zone and revolves it for a predetermined amount of time [21–26]. Friction lap welding (FLW), for instance, creates the junction by applying pressure and heat with the aid of a non-consumable circular revolving device. On top of the polymer is the metal sheet. To achieve the chosen pressure, the instrument is pushed as opposed to the metallic piece, and it is then rotated in the direction of welding [27]. The polymeric piece area next to the metal is melted by the heat produced by friction between the metal and the device, forming the junction with the pressed metal. FLW results in robust joints. However, in addition to uneven heat circulation along the weld line, it is only capable of producing overlapping joins [27].

4.2 LASER BEAM-BASED DEPOSITION

4.2.1 Laser beam powder bed fusion

One of the oldest metal additive manufacturing (MAM) approaches, laser beam powder bed fusion (LBPBF), selectively melts and consolidates powder into solid structures using a laser beam thermal energy source, as shown in Figure 4.1. Contingent on whether the metal is reactive or not, argon- or nitrogen-filled controlled environments are employed with reflective mirrors to transport the laser beam through a predetermined 2D examining route [28]. A substantial amount of research has been done on the final

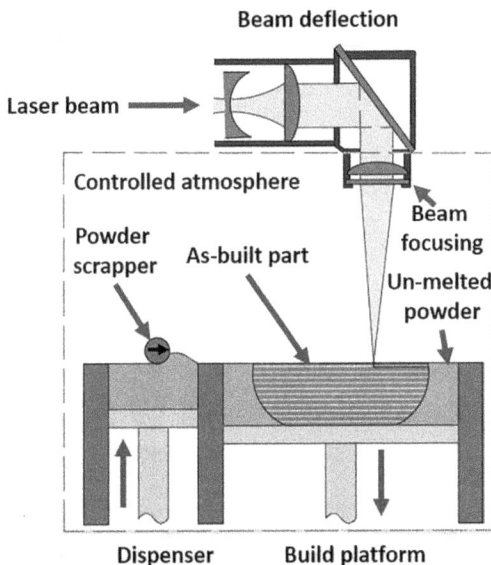

Figure 4.1 Schematic of the LBPBF process.

qualities of parts created using a wide variety of metal alloys using the mature MAM method known as LBPBF [29]. Due to the extensive usage of this method and equipment manufacturers' ongoing efforts to improve their products, it is currently possible to achieve accumulation amounts of up to 0.1 kg/h and surface roughness on the 10–20 scale. This explains the recent exponential surge in the sales of LBPBF equipment [30].

4.2.2 Laser beam direct energy deposition

Building three-dimensional metal parts layer by layer using the concepts of laser cladding is possible using laser beam direct energy deposition (LBDED), as shown in Figure 4.2. The feedstock might be any cable (wire) or powder; the latter is fed through the laser head (often coaxially) after being segmented, whereas the former is fed through a separate system from the latter. When employing wire as the feedstock, the maximum deposition rates can be as high as 2 kg/h, and the surface roughness is often better than 30 m. Since controlled environmental chambers are not necessary, the procedure is perhaps automated to increase track movement pliability for the creation of intricate 3D portions. In addition to shielding the melt pool from oxidation, shielding gases that flow from the laser head also help transfer powder to the melt pool by serving as carriers.

4.3 ELECTRON BEAM-BASED DEPOSITION

4.3.1 Electron beam powder bed fusion

The main difference between LBPBF and electron beam powder bed fusion (EBPBF) is that the heat energy required to melt the powder in EBPBF is

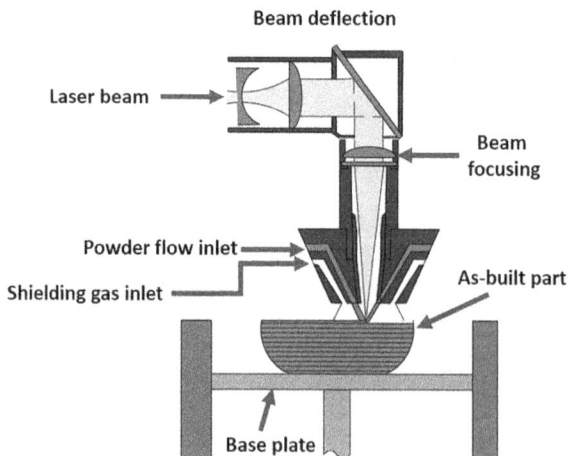

Figure 4.2 Schematic of the LBDED process.

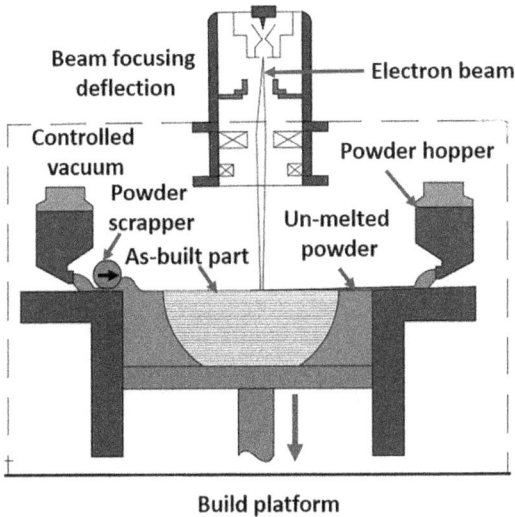

Build platform

Figure 4.3 Schematic of the EBPBF process.

from an electron beam rather than a laser beam, as shown in Figure 4.3. In addition to the change in the thermal energy source, the equipment has also undergone a number of additional modifications, including the electron beam's focus and deflection, utilizing electromagnetic lenses rather than mirrors. The feedstock needs to be preheated between 0.5 and 0.6 of its melting point in order to prevent powder dispersion brought on by electrostatic charging, or the so-called "powder pushed away phenomenon" [31]. EBPBF is used less frequently than LBPBF in both investigations and commercials. EBPBF has covered a number of industrial applications despite the fact that it is difficult to process materials [32]. The highest accumulation amounts of EBPBF, which are somewhat superior to LBPBF, can reach values of up to 0.2 kg/h with surface roughness in the range of 15–30 m.

4.3.2 Electron beam direct energy deposition

The operation of electron beam direct energy deposition (), as depicted in Figure 4.4, is like that of LBDED, with the exception of substituting an electron beam operating in a controlled temperature environment as the thermal energy. The technique only uses wire as a feedstock since insufficient treatment of metal powder movement might impact the superiority and correctness of the components. With deposition speed values ranging from 3 to 10 kg/h, depending on the material and part characteristics, EBDED makes it possible to fabricate large-scale parts. Large melt pools and high deposition rates result in severe thermal stresses that sometimes call for consideration of the substrate and fixture. The need for further procedures to be carried out in order to produce the finished parts renders the surface roughness of the constructed pieces meaningless.

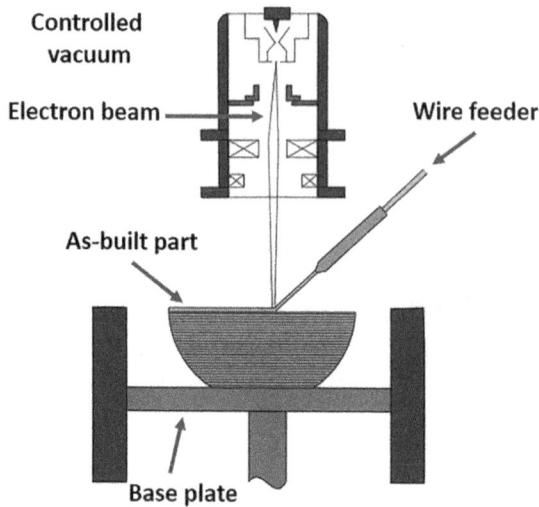

Figure 4.4 Schematic of the EBDED process.

4.4 METAL ARC-BASED DEPOSITION

4.4.1 Gas metal arc direct energy deposition

Gas metal arc welding, also called gas metal arc direct energy deposition (GMADED), is the most extensively used wire arc additive manufacturing (WAAM) technology. In this process, a consumable wire acts as an electrode, and it is automatically fed via a nozzle into the joint portion while being shielded by inert or active shielding gases, as shown in Figure 4.5 [33]. This GMADED process is an easy and cheap technique to use due to the wire feeding being coaxial with the nozzle.

4.4.2 Gas tungsten arc direct energy deposition and plasma arc direct energy deposition

The other two WAAM-based procedures depicted in Figures 4.6 and 4.7 are gas tungsten arc direct energy deposition (GTADED) and plasma arc direct energy deposition (PADED) [34,35]. In these two processes, electric arcs are produced by a non-consumable electrode (tungsten) and the metal component. The experimental mechanism of the GTADED technique is derived from gas tungsten arc welding equipment. Similarly, the experimental procedures of the PADED are derived from plasma arc welding equipment. In GMADED, a separate wire-feeding mechanism is used to deliver the wire relatively more than the nozzle [36]. In PADED, the accumulation amounts are superior to those in GTADED due to the electric arc features such as good stability, minimum thermal alteration, and large energy concentration [37].

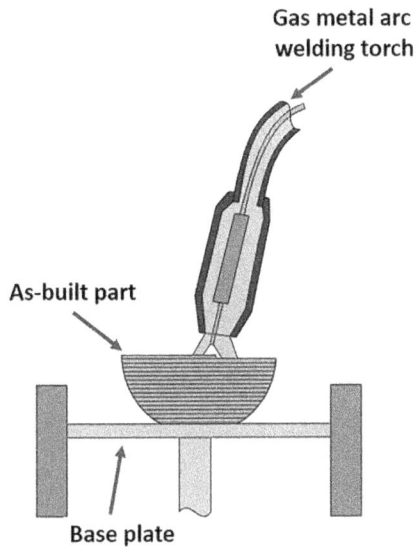

Figure 4.5 Schematic of the GMADED process.

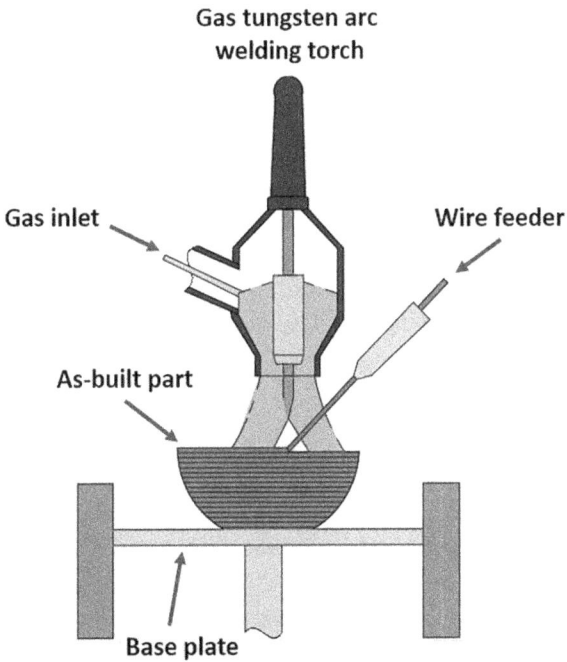

Figure 4.6 Schematic of the GTADED process.

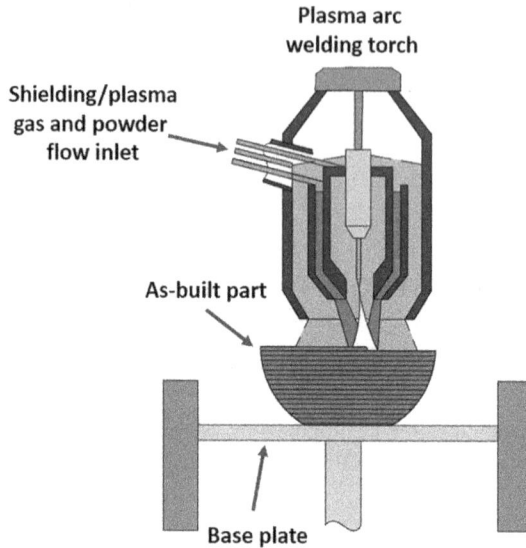

Figure 4.7 Schematic of the PADED process.

4.5 FRICTION STIR PROCESSING OF MMCS

Nowadays, academic and business institutions are looking for new and environmentally safe techniques for welding plastics to metals, which generate better weld quality and meet consumer demand. Friction stir welding (FSW) is one of the most recent, novel, and promising joining techniques in the scientific and industrial sectors [38]. It is described as a solid-state welding procedure that joins two neighbouring pieces using a rotating, non-consumable tool. The tool is made up of a shoulder and a pin. The shoulder of the pin is inserted into the neighbouring profile of the fastened pieces until it contacts their surface. The tool then turns and moves at a set speed in the welding direction. In parallel, joints are formed in the direction of the weld due to the production of frictional heat and material softening [39].

4.5.1 Tool parameters

In the FSW technique, the tool geometry is the most crucial component, and it is used to control the flow of material during the process [40]. The diameter of the shoulder, the diameter of the pin, and the shape and length of the pin are the important geometric tools of the FSW [41]. The tool is mostly made of steel and has two major functions: to facilitate material flow and localized heating. Friction between the pin and the workpiece is the core cause of the tool plunge heating up initially. Additionally, the material's distortion results in some additional heating [40]. Up until its shoulder makes contact with the surface of the material, the tool is pushed into the workpiece. The friction between the shoulder and the sample is the main heat source. The heated material's volume is also accommodated by the shoulder. Concave shoulders are the most common style of shoulder design.

The concave shape is made by creating a slight angle between the shoulder's edge and the pin, which makes it easy and simple to manufacture. In this type of shoulder, the angle of the tool should be inclined 2–4° away from the welding line's normal and in the direction of the weld journey [42]. Features on shoulders may be machined onto the shoulder profile of any tool. By directing worked material from the shoulder's edge to the pin, these properties improve material flow and eliminate the requirement for tool tilting [42]. The use of scrolled and grooved shoulder features to improve the quality of the joint in hybrid metal-polymer combinations has been documented in the literature [43,44]. High-quality joints are produced as a result of improved material flow and increased heat input from friction caused by large shoulder diameters. On the other hand, the small size of the shoulder affects the stir zone (SZ) of the material due to the flow of material and insufficient heat input [41].

Also, the geometry of the pin and shoulder heating is essential for heating. Additionally, tool pin shape impacts how the joint looks and dictates the uniformity of the microstructure, two crucial aspects of a high-quality weld [45]. Cylindrical threaded and tapered pin shapes were most frequently employed to effectively weld metals to plastics and produce joints that were relatively strong [43,44,46–48]. Huang et al. came to the conclusion that more pulsing movement might be produced by employing a tapered thread pin with triple facets. By applying the friction stir lap welding (FSLW) method, the flow of material is enhanced in comparison to a thread-tapered pin, and a joint quality of 20.6% of the plastic's strength is obtained [49]. The weld look is also influenced by the pin's length and diameter. A rough surface and significant valley-like flaws may result at the junction location if a big pin size fails to induce the plastic flow [50].

4.5.2 Properties of hybrid joints

i. Tensile strength

Numerous investigations were focused on the impact of various process feature changes on the weld quality [43,44,49,51–54]. The literature was used to regulate the tensile strength of metal-polymer hybrid joints at various process settings [55]. At 500 rpm and 40 mm/min, the highest tensile strength ever measured was 14.9 MPa. It was shown that even a small variation in feed rate significantly reduced the tensile strength. As a result, it may be said that FSW is a variable-sensitive method. Joint strength is found to grow with rotational speed and subsequently decline after a definite position, according to several studies in the literature [48,52,55].

ii. Microhardness

The hybrid joints are concaved in several places to assess the toughness of the welded joints. Researchers observed the shortest value of hardness for melted-resolidified polymers in the SZ region [46,47,49,52,54], which was probably caused by the polymer losing molecular weight and losing crystallinity. But, the results of the hardness in the SZ are higher compared to the base polymer due to the embedded metal pieces at the interface of the metal and polymer [47,49]. Since the metal pieces were exposed to significant strains before being cut with the FSW tool, they possessed a fine-grain structure and a high degree of toughness. The parameters like the rotational speed of the tool and its transverse rate affect the geometry and quantity of metal bits at the SZ, which could also have an impact on the hardness of the final component [51]. The recovery process causes the hardness at the SZ, thermos mechanically affected zone, and heat-affected zone (HAZ) sections on the metal side of the joint to diminish [54]. In the FSW method, the metal undergoes softening and reduces its joint strength through the application of heat. Also, in the HAZ, the value of hardness is lower compared to the SZ due to the production of very fine grains in the SZ [54].

iii. Thermal studies

According to experimental findings, friction plays a major role in producing a high amount of heat in the FSW technique [53]. Due to the heat, stress, and flow of the material, they are impacted in the FSW [56]. In order to deliver the required heat input, the variables used in the FSW should be tuned. Therefore, it is crucial to look at the joint thermal profile. The temperature in the weld portion quickly rises to a point that is briefly higher than the temperature at which the polymer is breaking down. The limited heat transfer coefficient of the polymer causes the subsequent cooling step to start slowly, keeping the temperature fields at the joint portion high [52]. The material undergoes local chemical and physical changes as a result of the highest temperature, which is always higher than the melting point and occasionally surpasses the polymer's thermal breakdown point [51–53]. The steady-state peak temperature time, however, is only a matter of seconds. Degradation of the polymer is therefore unlikely.

Additionally, increasing traverse speed lowers the FSW process's peak temperature and speeds up cooling, both of which improve joint characteristics. Temperatures rise as rotational speed increases because more heat is produced. Also, tilt angle and plunge depth were reported to have a similar pattern of behaviour in the FSW [51]. Furthermore, it is predicted that the peak interfacial temperature of aluminium-polymer junctions is somewhere between 0.5 and 0.6 of the melting point of aluminium alloys. Additionally, it is predicted that the peak interfacial temperature of aluminium-polymer junctions is somewhere between 0.5 and 0.6 of the melting point of aluminium alloys. The risk of thermal deterioration may be reduced by selecting a suitable thermal-compatible material.

4.6 CONCLUSION

The goal of this chapter began with an explanation of the need for hybrid metal-to-plastic joints and various methods for joining metal to plastic. This chapter is focused on the theory and applications of laser beam-based deposition, electron beam-based deposition, metal arc-based deposition, and friction stir processing. Additive manufacturing of polymer matrix composites and metal matrix composites is further explored. Besides, there is a discussion on the challenges that need to be overcome before these techniques can be widely adopted. The geometry of tool parameters and their effects during the FSW process are discussed. Moreover, the properties such as tensile strength, hardness, and thermal behaviour of hybrid joints in the FSW are discussed.

REFERENCES

[1] Sahu, S. K., Pal, K., Das, S. (2020), Parametric study on joint quality in friction stir welding of polycarbonate, *Materials Today: Proceedings*, 39, 1275–1280.
[2] Arif, M., Kumar, D., Siddiquee, A. N. (2022), Friction stir welding and friction stir spot welding of polymethyl methacrylate (PMMA) to other materials: areview, *Materials Today: Proceedings*, 62, 220–225.
[3] Kah, P., Suoranta, R., Martikainen, J., Magnus, C. (2014), Techniques for joining dissimilar materials: metals and polymers, *Review on Advanced Materials Science*, 36, 152–164.
[4] Wang, Z., Li, C., Sui, L., Xian, G. (2021), Effects of adhesive property and thickness on the bond performance between carbon fiber reinforced polymer laminate and steel, *Thin-Walled Structures*, 158, 107176.
[5] Heide-Jørgensen, S., Møller, R. K., Buhl, K. B., Pedersen, S. U., Daasbjerg, K., Hinge, M., Budzik, M. K. (2018), Efficient bonding of ethylene-propylene-diene M-class rubber to stainless steel using polymer brushes as a nanoscale adhesive, *International Journal of Adhesion and Adhesives*, 87, 31–41.
[6] Han, G., Tan, B., Cheng, F., Wang, B., Leong, Y. K., Hu, X. (2021), CNT toughened aluminium and CFRP interface for strong adhesive bonding, *Nano Materials Science*, 4, 266–275.
[7] Cheng, F., Hu, Y., Zhang, X., Hu, X., Huang, Z. (2021), Adhesive bond strength enhancing between carbon fiber reinforced polymer and aluminum substrates with different surface morphologies created by three sulfuric acid solutions, *Compos, Composites Part A: Applied Science and Manufacturing*, 146, 106427.
[8] Heydari, M., Sharif, F., Ebrahimi, M. (2021), A molecular dynamics study on the role of oxygen-containing functional groups on the adhesion of polymeric films to the aluminum surface, *Fluid Phase Equilibria*, 536, 112966.

[9] Wang, J., Zhang, G., Zheng, X., Li, J., Li, X., Zhu, W., Yanagimoto, J. (2020), A self-piercing riveting method for joining of continuous carbon fiber reinforced composite and aluminum alloy sheets, *Composite Structures*, 259, 113219.

[10] Vignesh, N. J. (2021), Analytical approach for modelling of heat generation in low-speed friction riveting of polymer/aluminium joints, *Materials Today: Proceedings*, 47, 6835–6838.

[11] Hynes, N. R. J., Sankaranarayanan, R., Sujana, J. A. J. (2021), A decision tree approach for energy efficient friction riveting of polymer/metal multi-material lightweight structures, *Journal of Cleaner Production*, 292, 125317.

[12] Gay, A., Lefebvre, F., Bergamo, S., Valiorgue, F., Chalandon, P., Michel, P., Bertrand, P. (2015), Fatigue of aluminum/glass fiber reinforced polymer composite assembly joined by self-piercing riveting, *Procedia Engineering*, 133, 501–507.

[13] Staab, F., Liesegang, M., Balle, F. (2020), Local shear strength distribution of ultrasonically welded hybrid Aluminium to CFRP joints, *Composite Structures*, 248, 112481.

[14] Staab, F., Balle, F. (2019), Ultrasonic torsion welding of ageing-resistant Al/CFRP joints: properties, microstructure and joint formation, *Ultrasonics*, 93, 139–144.

[15] Mongan, P. G., Hinchy, E. P., ODowd, N. P., McCarthy, C. T. (2021), Quality prediction of ultrasonically welded joints using a hybrid machine learning model, *Journal of Manufacturing Processes*, 71, 571–579.

[16] Katayama, S., Kawahito, Y. (2008), Laser direct joining of metal and plastic, *Scripta Materialia*, 59, 1247–1250.

[17] Zhou, M., Xiong, X., Drummer, D., Jiang, B. (2019), Interfacial interaction and joining property of direct injection-molded polymer-metal hybrid structures: a molecular dynamics simulation study, *Applied Surface Science*, 478, 680–689.

[18] Zhao, S., Kimura, F., Wang, S., Kajihara, Y. (2021), Chemical interaction at the interface of metal-plastic direct joints fabricated via injection molded direct joining, *Applied Surface Science*, 540, 148339.

[19] Wang, S., Kimura, F., Zhao, S., Yamaguchi, E., Ito, Y., Kajihara, Y. (2021), Influence of fluidity improver on metal-polymer direct joining via injection molding, *Precision Engineering*, 72, 620–626.

[20] Bonpain, B., Stommel, M. (2018), Influence of surface roughness on the shear strength of direct injection molded plastic-aluminum hybrid-parts, *International Journal of Adhesion and Adhesives*, 82, 290–298.

[21] Lambiase, F., Grossi, V., Paoletti, A. (2022), High-speed joining of hybrid metal-polymer joints during the friction-assisted joining process, *Composite Structures*, 280, 114890.

[22] Lambiase, F., Paoletti, A., Durante, M. (2021), Mechanism of bonding of AA7075 aluminum alloy and CFRP during friction assisted joining, *Composite Structures*, 261, 113593.

[23] Lambiase, F., Balle, F., Blaga, L. A., Liu, F., Amancio-Filho, S. T. (2021), Friction-based processes for hybrid multi-material joining, *Composite Structures*, 266, 113828.

[24] Lambiase, F., Grossi, V., Paoletti, A. (2021), Defects formation during Friction Assisted Joining of metals and semi crystalline polymers, *Journal of Manufacturing Processes*, 62, 833–844.

[25] Lambiase, F., Paoletti, A. (2018), Mechanical ior of AA5053/polyetherether-ketone (PEEK) made by friction assisted joining, *Composite Structures*, 189, 70–78.

[26] Lambiase, F., Paoletti, A., Grossi, V., Di Ilio, A. (2017), Friction assisted joining of aluminum and PVC sheets, *Journal of Manufacturing Processes*, 29, 221–231.

[27] Liu, F. C., Liao, J., Nakata, K. (2014), Joining of metal to plastic using friction lap welding, *Materials & Design*, 54, 236–244.

[28] Pragana, J. P., Pombinha, P., Duarte, V. R., Rodrigues, T. A., Oliveira, J. P., Bragança, I. M., Santos, T. G., Miranda, R. M., Coutinho, L., Silva, C. M. (2020), Influence of processing parameters on the density of 316L stainless steel parts manufactured through laser powder bed fusion, *Proceedings of the Institution of Mechanical Engineers, Part B: Journal of Engineering Manufacture*, 234 (9), 1246–1257.

[29] Bhavar, V., Kattire, P., Patil, V., Khot, S., Gujar, K., Singh, R. (2014), A review on powder bed fusion technology of metal additive manufacturing, 4th International conference and exhibition on Additive Manufacturing Technologies-AM-2014, Banglore, India.

[30] Wohlers, T. (2017), Desktop metal: a rising star of metal am targets speed, cost and high-volume production, *Metal AM*, 3.

[31] Murr, L. E., Martinez, E., Amato, K. N., Gaytan, S. M., Hernandez, J., Ramirez, D. A., Shindo, P. W., Medina, F., Wicker, R. B. (2012), Fabrication of metal and alloy components by additive manufacturing: examples of 3D materials science, *Journal of Materials Research and Technology*, 1, 42–54.

[32] Korner, C. (2016). Additive manufacturing of metallic components by selective electron beam melting-a review, *International Materials Reviews*, 61 (5), 361–377.

[33] Williams, S. W., Martina, F., Addison, A. C., Ding, J., Pardal, G., Colegrove, P. (2016). Wire+arc additive manufacturing, *Materials Science and Technology*, 32, 641–647.

[34] Baufeld, B., Van der Biest, O., Gault, R. (2010), Additive manufacturing of Ti-6Al-4V components by shaped metal deposition: microstructure and mechanical properties, *Materials & Design*, 31, 106–111.

[35] Martina, F., Mehnen, J., Williams, S. W., Colegrove, P., Wang, F. (2012), Investigation of the benefits of plasma deposition for the additive layer manufacture of Ti-6Al-4V, *Journal of Materials Processing Technology*, 212, 1377–1386.

[36] Wu, B., Pan, Z., Ding, D., Cuiuri, D., Li, H., Xu, J., Norrish, J. (2018), A review of the wire arc additive manufacturing of metals: properties, defects and quality improvement, *Journal of Manufacturing Processes*, 35, 127–139.

[37] Zhang, H., Xu, J., Wang, G. (2003), Fundamental study on plasma deposition manufacturing, *Surface and Coatings Technology*, 171, 112–118.

[38] Braga, D. F. O., Eslami, S., Moreira, P. M. G. P., Da Silva, L., El-Zein, M., Martins, P. (2021), Friction stir welding. In *Advanced Joining Processes*; Elsevier: Amsterdam, The Netherlands, 173–206.

[39] El-Sayed, M. M., Shash, A., Abd-Rabou, M., ElSherbiny, M. G. (2021), Welding and processing of metallic materials by using friction stir technique: a review, *Journal of Advanced Joining Processes*, 3, 100059.

[40] Mishra, R. S., Ma, Z. Y. (2005), Friction stir welding and processing, *Materials Science & Engineering R: Reports*, 50, 1–78.
[41] Huang, Y., Meng, X., Xie, Y., Wan, L., Lv, Z., Cao, J., Feng, J. (2018), Friction stir welding/processing of polymers and polymer matrix composites, *Composites Part A: Applied Science and Manufacturing*, 105, 235–257.
[42] Meilinger, A., Török, I. (2013), The importance of friction stir welding tool, *Production Processes and Systems*, 6, 25–34.
[43] Wang, T., Li, L., Pallaka, M. R., Das, H., Whalen, S., Soulami, A., Upadhyay, P., Kappagantula, K. S. (2021), Mechanical and microstructural characterization of AZ31 magnesium-carbon fiber reinforced polymer joint obtained by friction stir interlocking technique, *Materials and Design*, 198, 109305.
[44] Ratanathavorn, W., Melander, A. (2015), Dissimilar joining between aluminium alloy (AA 6111) and thermoplastics using friction stir welding, *Science and Technology of Welding and Joining*, 20, 222–228.
[45] Payganeh, G. H., Arab, N. B. M., Asl, Y. D., Ghasemi, F. A., Boroujeni, M. S. (2011), Effects of friction stir welding process parameters on appearance and strength of polypropylene composite welds, *International Journal of the Physical Sciences*, 6, 4595–4601.
[46] MirHashemi, S., Amadeh, A., Khodabakhshi, F. (2021), Effects of SiC nanoparticles on the dissimilar friction stirweldability of low-density polyethylene (LDPE) and AA7075 aluminum alloy, *Journal of Materials Research and Technology*, 13, 449–462.
[47] Khodabakhshi, F., Haghshenas, M., Sahraeinejad, S., Chen, J., Shalchi, B., Li, J., Gerlich, A. (2014), Microstructure-property characterization of a friction-stir welded joint between AA5059 aluminum alloy and high density polyethylene, *Materials Characterization*, 98, 73–82.
[48] Dalwadi, C. G., Patel, A. R., Kapopara, J. M., Kotadiya, D. J., Patel, N. D., Rana, H. (2018), Examination of mechanical properties for dissimilar friction stir welded joint of Al alloy (AA-6061) to PMMA (Acrylic), *Materials Today: Proceedings*, 5, 4761–4765.
[49] Huang, Y., Meng, X., Wang, Y., Xie, Y., Zhou, L. (2018), Joining of aluminum alloy and polymer via friction stir lap welding, *Journal of Materials Processing Technology*, 257, 148–154.
[50] Rahmat, S. M., Hamdi, M., Yusof, F., Moshwan, R. (2014), Preliminary study on the feasibility of friction stir welding in 7075 aluminium alloy and polycarbonate sheet, *Materials Research Innovations*, 18, 515–519.
[51] Derazkola, H.A., Simchi, A. (2019), An investigation on the dissimilar friction stir welding of T-joints between AA5754 aluminum alloy and poly(methyl methacrylate), *Thin-Walled Structures*, 135, 376–384.
[52] Derazkola, H. A., Elyasi, M. (2018), The influence of process parameters in friction stir welding of Al-Mg alloy and polycarbonate, *Journal of Manufacturing Processes*, 35, 88–98.
[53] Derazkola, H. A., Simchi, A. (2020), A new procedure for the fabrication of dissimilar joints through injection of colloidal nanoparticles during friction stir processing: proof concept for AA6062/PMMA joints, *Journal of Manufacturing Processes*, 49, 335–343.

[54] Shahmiri, H., Movahedi, M., Kokabi, A. H. (2017), Friction stir lap joining of aluminium alloy to polypropylene sheets, *Science and Technology of Welding and Joining*, 22, 120–126.

[55] Patel, A. R., Kotadiya, D. J., Kapopara, J. M., Dalwadi, C. G., Patel, N. P., Rana, H. (2018), Investigation of mechanical properties for hybrid joint of aluminium to polymer using friction stir welding (FSW). *Materials Today: Proceedings*, 5, 4242–4249.

[56] Li, M., Xiong, X., Ji, S., Hu, W., Yue, Y. (2021), Achieving high-quality metal to polymer-matrix composites joint via top-thermic solid-state lap joining, *Composites Part B: Engineering*, 219, 108941.

Chapter 5

A review on heat treatments, microstructure, and mechanical properties of selective laser-melted AISI 316L

Tushar R. Dandekar
University of Portsmouth

5.1 INTRODUCTION

5.1.1 Introduction to selective laser melting

Additive manufacturing (AM), as defined by the ISO/ASTM 52900 terminology standard, is the process of joining materials to make parts from 3D model data. Usually, the material is joined layer upon layer, as opposed to subtractive and formative methods of manufacturing. Over time, AM has been extensively applied to a variety of metallic systems, particularly Al alloys, Ni-based superalloys, steels, and Ti alloys [1,2]. Based on the ISO/ASTM 52900 terminology standard, the two basic techniques for metal AM are direct-energy deposition and powder-bed fusion. While the former allows for direct reconstruction and repair of degraded areas and is a speedier process, only the latter allows for the fabrication of complicated forms and produces a finer and less porous microstructure. A 3D item is generated layer by layer from a metal powder feedstock that is selectively melted by a high-energy laser source or electron beam in the powder-bed fusion manufacturing technique. Following cooling, the generated component is consolidated. This procedure is known by various business names, such as selective laser melting (SLM), commonly known as laser powder bed fusion or direct metal laser melting, which is an AM process designed to melt and fuse metallic powders using a high power density laser [3,4]. The SLM technique works by applying extremely thin layers of metallic powder to a building platform, which are entirely melted subsequently by the thermal energy produced by one or more laser beams. By carefully choosing which metallic powders to selectively melt and then re-solidify in each layer, the desired 3D part's cross-section area is constructed. A new coating of powders is then applied and levelled by a re-coater after the building platform has been slightly lowered.

Specifically, designed scanner optics can guide and concentrate the laser beam(s) through a computer-generated pattern. In order to construct 3D

DOI: 10.1201/9781003406488-5

things based on the CAD design, the powder particles may be selectively melted in the powder bed. A variety of metal powders, including those made of Al, Ti, Cu, Cr, Co, stainless steel, tool steel, and superalloy, have been successfully used in SLM processes. Although the majority of discarded powders could be recycled for subsequent AM processes, it is challenging and inefficient to fill the build volume in the SLM working chamber, particularly when large components are needed. Some material loss happens when the powders become polluted or oxidized during the melting process, rendering them unrecyclable [5–7].

When it comes to producing geometries with arbitrary designs, SLM has several limitations. Due to inadequate heat conduction in the powder bed just beneath the freshly hardened layers of exposed powders, overhanging forms or horizontal struts are currently challenging to fabricate. The majority of commercial SLM systems use powders with particle sizes ranging from 20 to 50 μm and a typical layer thickness of 20–100 μm. The advancement of traditional SLM scaling for enhanced feature resolution is primarily focused on three factors: powder particle size, laser beam diameter, and layer thickness. In micro-SLM systems, both continuous-wave and pulsed lasers have been used [3].

5.1.2 Introduction to austenitic stainless steels

Because of their exceptional corrosion resistance, biocompatibility, and ductility, austenitic stainless steels (ASSs) are one of the most used groups of industrial alloys. These properties make them ideal candidates for applications in a variety of industry segments, including biomedical, aerospace, defence, oil & gas, petrochemical, and more [4,8,9]. The most popular category of steel in AM is ASSs, particularly 316L grade. While there are still numerous difficulties, AM has been found to have some promising features in this case. The literature is replete with publications on AM and ASSs. It would take too much time to go over each one separately in this part. As a result, we shall examine the trends in microstructure evolution, mechanical properties, and deformation mechanisms, together with their advancements and difficulties. It should be emphasized that the majority of the literature on AM of ASSs concentrates on grade 316L, which is the most often used material in many industrial applications. Other ASSs, such as the 304L grade, are the second most reported grade. The most significant difference between 316L and 304L is in their chemical compositions, with 316L containing almost 2% wt% Mo to increase corrosion resistance. In general, a lot of work has gone into examining the mechanical properties of AM metals and alloys. Numerous articles have reviewed the processing-microstructure properties of AM metals, including their mechanical characteristics [1,4,10,11]. The three most thorough evaluations of AM steels that are currently accessible are those by Fayazfar et al. [12], Bajaj et al. [8], and Haghdadi et al. [2]. In-depth summaries of the processing of steels using several powder-based AM systems were provided by Fayazfar et al. [12]. These authors reviewed the AM steels' solidification microstructure and went through the fundamentals of the powder bed, powder-fed, and binder jetting processes. The development of steel microstructures during AM was reviewed by Bajaj et al. [8]. These authors provided a summary of the various steel series used in AM industrial operations and analysed the link between microstructure and properties in these steels. In addition, Haghdadi et al. [2] discussed the achievements and challenges associated with the AM of steels.

Despite significant progress in understanding the unique microstructural features and their effect on the mechanical properties of AM low-strength to high-strength alloys compared to conventionally treated alloys, there is a dearth of knowledge on how they can potentially result in improved attributes. The current study aims to close this gap by giving a complete assessment of the unique features of AM alloys that have been documented in the literature. Thus, the current book chapter aims to discuss the effect of heat treatments (HTs) on the mechanical properties of SLMed additively manufactured ASS (316L SS). Thus, we largely focused on the microstructure evolution, mechanical properties, and associated deformation mechanism of SLMed additively manufactured ASS (316L SS), where appropriate investigations are available. The abbreviation SLMed AISI 316L is used hereafter for selective laser-melted 316L ASS.

5.2 MICROSTRUCTURE EVOLUTION DURING SLM

The microstructure of a material has a substantial influence on its performance [8,12]. To find the morphological characteristics and distributions of constituent phases occurring inside the microstructure, several characterization technologies, such as optical microscopy (OM), scanning electron microscopy (SEM), and transmission electron microscopy (TEM), were employed. As a consequence, the microstructure explanation discloses a slew of characteristics such as grain structure (size, shape, and distribution), interfaces (grain boundary and interphase boundary), the existence of distinct phases (size, shape, distribution, and volume percentage), and dislocation substructure (largely density and distribution). The crystallographic texture, which is frequently overlooked, is also an important component of the microstructure [8,13–15]. However, in this section, only microstructure evolution during SLM and after post-heat processing treatment is discussed in detail.

The type of fabrication process has been reported [16] to alter the microstructure as well as the mechanical performance of steels. Figure 5.1 [16] depicts a distinction between the microstructures of AM, as-cast, and forged AISI 316L specimens. As can be observed, the AM technique produces distinctive reversed-bell-like grains. The optical micrographs

(Figure 5.1, inset) clearly display the melt pools (MPs) formed by the SLM of the parent powders. After etching the specimen, the melt-pool boundaries (MPBs) are observable under the OM, indicating either a distinct crystal structure or the existence of compositional changes. The MPs are separated by 50–80 μm in the build plane (BP). This generally correlates to the laser-hatching distance. The conduction melting mode is basically responsible for the extended and shallow geometry of the MPs [17]. The base part of the grains corresponds to the geometry of the MPs, as shown by the EBSD IPF maps (Inverse Pole Figure (IPF), as the name suggests, are actually the reverse of the standard Pole Figures. Instead of plotting crystallographic directions in the sample reference frame, IPFs plot a sample direction in the crystallographic reference frame.), which were captured with the same magnification as the overlaid optical micrograph (Figure 5.1a). This suggests that the majority of the grains nucleate at the bottom part of the MPs. On the contrary, the grains proceed considerably further in the BP than the individual MPs do. The grains can grow to lengths of many hundreds of microns in this BP, which equates to a layer thickness of three to five. This suggests that by epitaxially growing on the pre-existing grains, the grains may spread over many deposition layers.

For the AM specimen, the average area grain size and grain diameter were found to be 347 μm^2 and 13.8 μm, respectively. Nevertheless, irregular, and anisotropic behaviour was observed for the grain shape. The grain-boundary density of the AM specimen was observed to be 1.5× higher than that of the forged specimen. In addition, twin boundaries (TBs) account for just ~1% of grain boundaries. On the other hand, forged and as-cast specimens showed an altogether altered microstructure. For forged specimens, the average area grain size was found to be 5,000 μm^2 (without considering twin TBs) and 1,000 μm^2 (with TBs). While specifically for equiaxed grains, the average grain size was 65 μm (without considering TBs) and 27 μm (with TBs). Approximately 56% of the high-angle grain boundaries (HAGB) are contributed by the TBs, with an HAGB density of 7.1×10^4 m^{-1}. No specific orientation (colour gradient) was shown by the forged specimen as observed

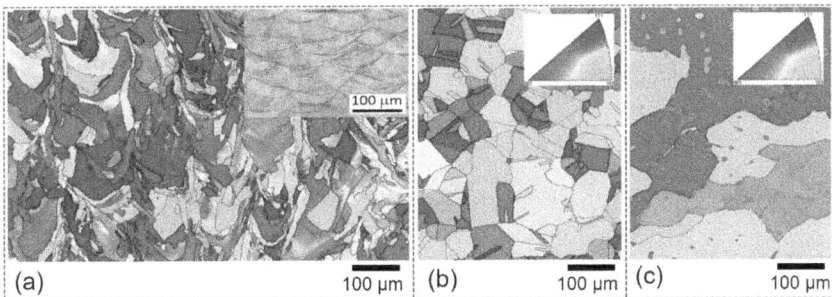

Figure 5.1 EBSD IPF maps of AISI 316L specimens processed through various routes: (a) AMed, (b) forged, and (c) as-cast [16] with permission from Elsevier.

by EBSD IPF maps. The as-cast steel microstructure (Figure 5.1c) is made up of incredibly big grains with a FCC (A face-centered cubic unit cell structure consists of atoms arranged in a cube where each corner of the cube has a fraction of an atom with six additional full atoms positioned at the center of each cube face.) crystal structure and tiny inclusions of BCC δ-ferrite. EBSD IPF maps likewise (forged specimen) did not reveal any specific orientations inside individual grains [16].

Figure 5.2 depicts the optical micrograph, secondary electron SEM micrograph, and EBSD IPF map in the BP of the AM specimen. The boundaries of the MP are shown by the black lines in Figure 5.2a and b. The differing etching behaviour of the black lines is most likely owing to the varying elements in the constitution of these regions. Figure 5.2b shows a cell-like structure (cellular dendritic structure) inside a single grain of order 0.5 μm. These distinctive structures were following MPB alongside the grains and were noticed across the entire AM specimen. Doubled MPBs were seen in a few regions, which were initiated due to metal repeatedly remelting and solidifying during successive layer deposition and laser interaction. According to reports [18,19] in the literature, there are no cell structures in the regions between these double MPBs. Conversely, in the discussed AM specimen, these structures could be seen throughout the whole substance and were uniformly dispersed [16]. Figure 5.2c demonstrates that the

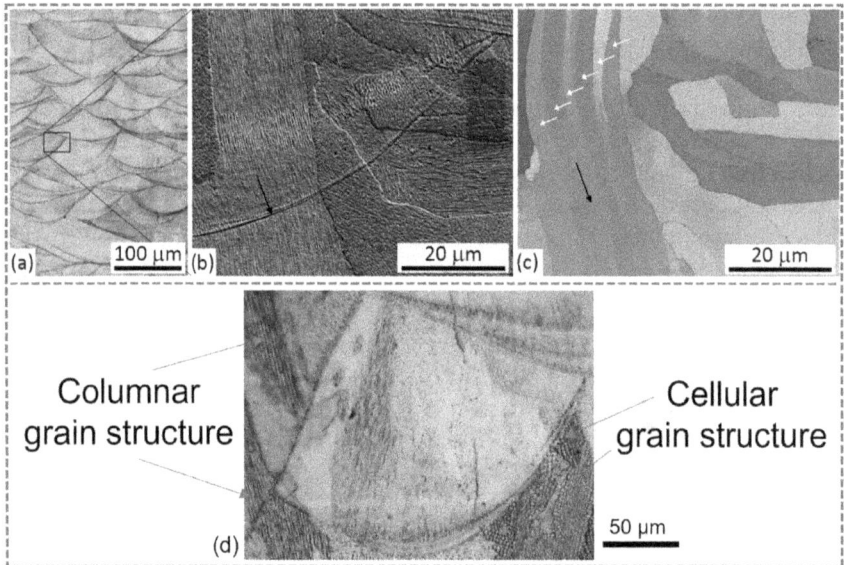

Figure 5.2 Melt-pool boundaries are represented by (a) optical micrograph, (b) secondary electron SEM micrograph, (c) EBSD IPF maps overlaid with an image quality map in BP [16] with permission from Elsevier, and (d) high magnification optical micrograph. (White arrows indicate an orientation gradient driven by relieving stress, and black arrows indicate MPB).

crystal orientations persist over the MPB with no measurable discontinuity in EBSD (which is represented by a black arrow in Figure 5.2c). Figure 5.2d shows a high magnification optical micrograph consisting of both cellular dendritic structures and columnar grain structures.

To enhance the mechanical properties, various types of HT can be applied. Ozer et al. [20] studied the effect of HTs and demonstrated the variation in microstructure as shown in Figure 5.3. Figure 5.3a shows the cellular dendritic structure and columnar grain structures of the as-built sample. These distinctive structures were following MPB along with the grains and were

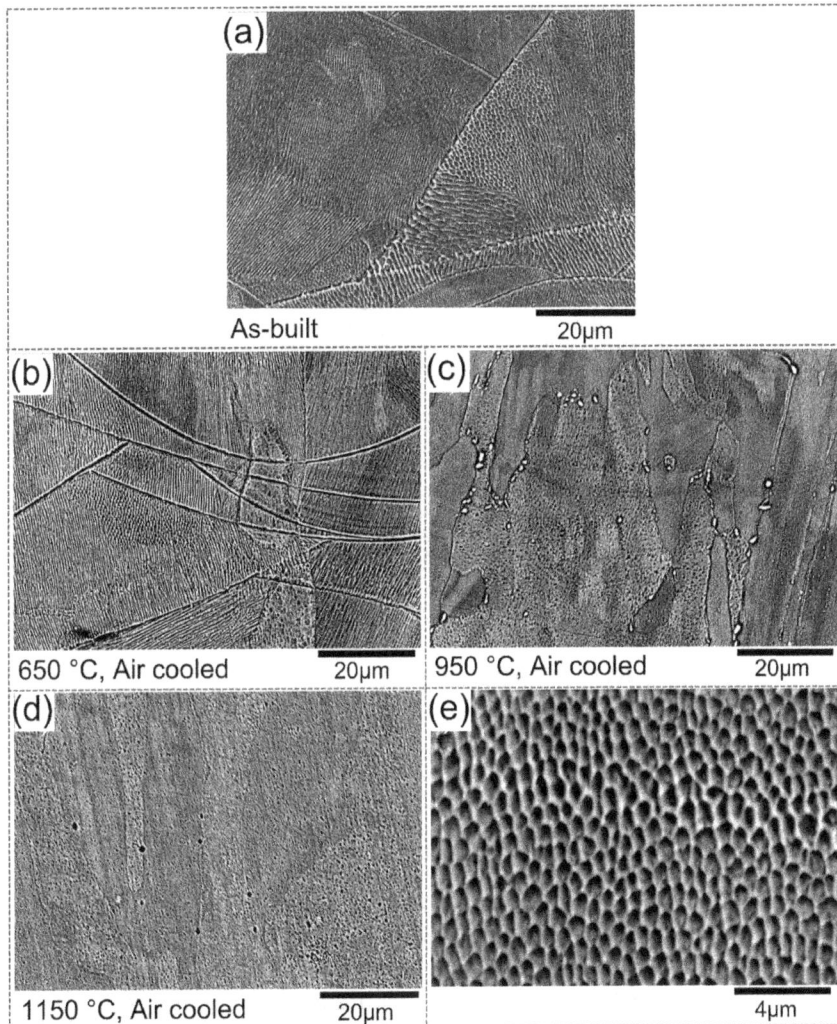

Figure 5.3 SEM micrographs of (a) as-built, (b, c, and d) different heat-treated samples and (e) SEM images of the MP on BP [20] with permission from Elsevier.

noticed across the entire AM specimen. Figure 5.3b–d shows the effect of HTs on the variation of microstructure. Lower temperatures did not show huge variation, while increases in temperature led to destruction of MPBs and grain coarsening. Similarly, Tascioglu and co-workers [21] inspected the effect of HT temperature on mechanical behaviour and carried out various HTs (namely HT-1, HT-2, and HT-3) at temperatures of 600°C, 850°C, and 1,100°C followed by 2 hours of soaking time and subsequent air cooling, respectively. The SEM micrograph demonstrates that the MPs and their boundaries with cellular and band structures were still discernible under the HT-1 condition, indicating a minimal alteration in the microstructure. This is mostly due to the low HT temperature in comparison to the sample melting point. The SEM micrograph clearly demonstrates that after the sample was treated to the HT-2 condition, the MPBs were almost imperceptible and lacked cellular features. In this instance, grain type morphology was formed along with newer grains, and the heat-treated sample microstructure clearly reveals the grain boundaries. From all the heat-treated conditions, the treatment of HT-3 was determined to have the greatest effect on the microstructure. Here, the MP and associated substructures, as well as its boundaries, totally vanished, and the entire sample consisted of newly formed recrystallized grains with coarse morphology. The sample was handled at a high temperature in the HT-3 condition, so coarser grains in HT-3 stand out when compared to HT-2. Although 316L is a single-phase FCC material, the martensite phase can be observed in the structure of the 316L as an outcome of HT and particular processing [6,9]. Due to the material's rapid solidification and the presence of chemical components that promote the production of ferrite with Si, Cr, and Mo, considerable ferrite development may also be seen in the microstructure. It is believed that the HT results in a drop-in ferrite density, which lowers the material's hardness [21].

5.3 MECHANICAL PROPERTIES

Post-processing HTs is crucial in establishing the microstructural features of the AM metallic alloys, and thus the mechanical properties of such alloys may undergo variation as a result of the microstructural changes. This section focuses on the mechanical properties of SLMed AISI 316L, including yield strength (YS), ultimate tensile strength (UTS), % elongation and hardness. Table 5.1 shows the mechanical properties of 316L ASS fabricated by SLM after various types of post-processing HTs reported in the literature. As it can be observed, there are a lot of investigations available in the reported literature; however, few important studies are discussed in this section.

Chen and co-workers [32] studied the influence of HT (400°C accompanied by water quenching) on mechanical properties and stated that YS

Table 5.1 Mechanical properties of 316L ASS fabricated by SLM after various types of post-processing heat treatments reported in literature

Condition	Orientation	YS (MPa)	UTS (MPa)	Elongation (%)	P (W)	v (mm/s)	Layer thickness (µm)	Hatch ratio (%)	ρ (%)	References
AF	Horizontal	590	700	58	150	700	30	~50	–	Wang et al. [18]
	Vertical	450	640	>70	296	150	50	–	–	
AF	Horizontal	557±14	519±12	42±2	200	–	50	–	99.9	Tolosa et al. [22]
	Vertical	602±47	664±47	30±47						
AF	Horizontal	517±7	634±7	33±0.6	200	800	30	–	–	Hitzler et al. [23]
	Vertical	439±10	512±18	12±5						
AF	Horizontal	642–643	517–745	15–28	90	–	50	–	99	Meier & Haberland [24]
	Vertical	549–589	595–636	14–18						
AF	Horizontal	444±27	567±19	8±2.9	175	700	60	–	93.8±2.6	Mertens et al. [25]
	Vertical	534±6	653±3	16.2±0.8					97.5±1	
AF	Horizontal	406±20	510±4	18±1	100	400	50	–	98	Rottger et al. [26]
	Vertical	427±8	522±5	15±2						
SLM+HIP densified 1150°C×3 hours	Horizontal	201±4	428±13	38±6						
AF	–	590±17	705±15	44±7	85	222 or 400	30	70	97.8	Carltons et al. [27]
SLM+solution annealing 1095°C×1 hour	–	375±11	635±17	51±3						
AF	Horizontal	638±11	752±16	41±3	195	1083	25	70	99.9	Kong et al. [28,29]
SLM+solution annealing 1,050°C×2 hours	Horizontal	424±8	673±13	44±3						
SLM+solution annealing 1,200°C×2 hours	Horizontal	416±9	684±16	52±3						
SLM+Stress-relief 388°C×4 hours	Horizontal	496	717	28	195	750	40	–	–	Mower and Long [30]
AF	Horizontal	640	760	30	103	425	30	–	–	Spierings et al. [31]

and micro-hardness (252 HV → 291 HV) increased while decreasing % elongation (~2%) due to application of HT. It was believed that newly generated silicate precipitation increased micro-hardness and yield strength. A further increase in HT temperature (800°C) leads to a decrease in YS and micro-hardness, as seen in Figure 5.4 [32]. Similar to this, it was discovered that the YS of SLMed AISI 316L that had undergone HT was 482 and 586 MPa, respectively, with nearly identical percentages of elongation. It has been suggested that the drop-in dislocation density (DD) inside the cell structure is what causes the reduction in YS and micro-hardness at this temperature. Likewise, HT from 1,050°C to 1,200°C decreased micro-hardness from 240 HV to 190 HV. Moreover, YS also decreased in the order of 637 MPa > 423 MPa > 415 MPa with a rise in HT temperature. The recrystallization phenomenon that took place at such high temperatures has the potential to be responsible for this. An alternative reason for this reduction in YS could be associated with grain coarsening due to an increase in HT temperature, which can be linked to the Hall-Petch relation. A cellular structure linked with SLMed AISI 316L enhanced the YS and micro-hardness in comparison with wrought AISI 316L [32,33].

In another study, Salman and co-workers [34] analysed the effect of HT on microstructure and mechanical properties and concluded that the strength of the AISI 316L samples reduced with an increase in annealing temperature due to microstructure coarsening. Additionally, the complex microstructure and subgrains were primarily responsible for the good ductility and superior strength (YS = 550 ± 10 MPa, UTS = 1,016 ± 8 MPa, and plastic deformation greater than 50%) of the as-SLM specimen compared to the annealed specimen, as shown in Figure 5.5 [34].

A similar study was conducted by Chao et al. [35] stated similarly, YS steadily decreased as the HT temperature increased over 400°C; however, conversely, the UTS was very minimally affected, as shown in Figure 5.6. As a result, the YS/UTS ratio decreased from around 70% in the as-built state

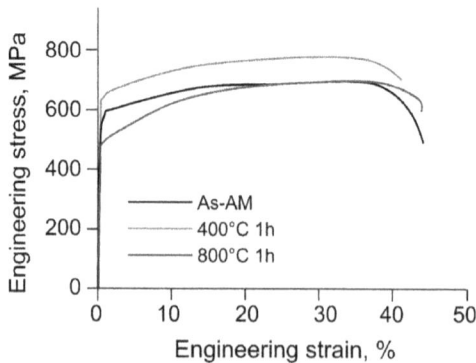

Figure 5.4 Engineering stress–strain curve of SLMed AISI 316L and heat-treated specimens at various temperatures [32] with permission from Elsevier.

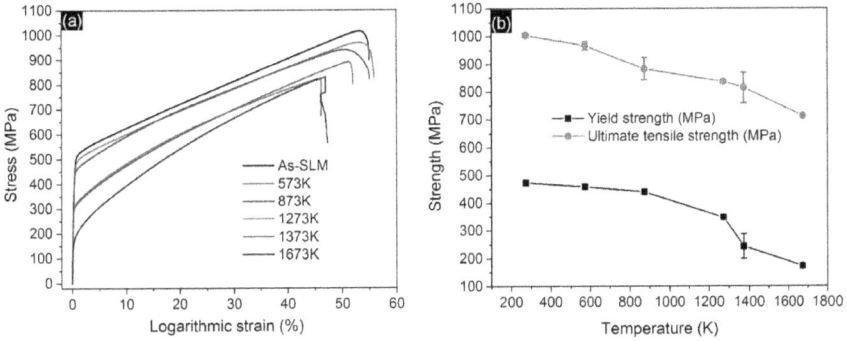

Figure 5.5 (a) Stress–strain curve for as-SLM and annealed specimens, (b) effect of HT on UTS and YS of annealed specimens [34] with permission from Elsevier.

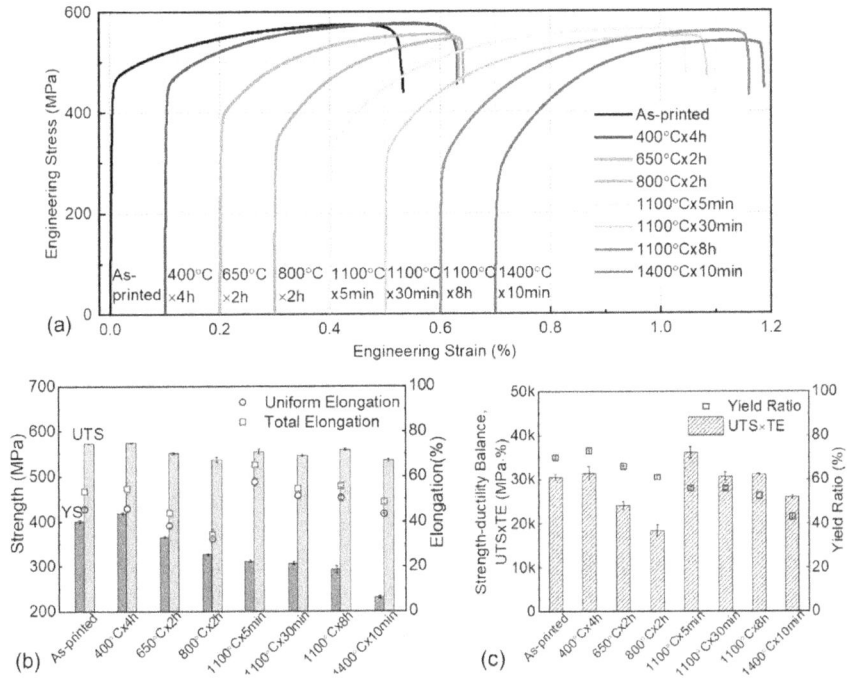

Figure 5.6 (a) Engineering stress–strain curve, (b) mechanical property metrics, and (c) strength-ductility product of as-printed and heat-treated SLMed AISI 316L [35] with permission from Elsevier.

to 43% after 10 minutes at 1,400°C (Figure 5.6c). As the HT temperature rose from 400 to 800°C (55%–34%), the total elongation (TE) at fracture decreased. The entire stress-relief treatment at 1,100°C for 5 minutes yielded the greatest TE of the investigated circumstances (65%), which is 12% higher than the as-built state (Figure 5.6b). After a long period of holding within the same temperature, the TE dropped to 55% and 58% after 30 minutes and 8 hours, respectively. A further rise in the HT temperature to 1,400°C sustained for 10 minutes was notably associated with a lower elongation of 49%. UTS×TE, which stands for the product of UTS and TE, represents the balance between the strength and ductility of steel for engineering purposes. UTS×TE significantly dropped from ~305 GPa.% for the as-built state to ~184 GPa.% following post-treatment at 800°C for 2 hours (Figure 5.6c).

5.4 DEFORMATION MECHANISMS

A common alloy utilized in parts with intricate internal structures for nuclear power plants and biomedical uses, SLMed AISI 316L often shows an extraordinary strength-ductility combination. One of the main factors contributing to these outstanding mechanical capabilities is the cellular substructure that results from the rapid solidification during SLM. High DD and the Hall-Petch effect of the cellular substructure have been specifically identified as the key causes of the high YS of SLMed AISI 316L.

The stacking fault energy (SFE) of AISI 316L that has undergone conventional processing is reported to be ~10–30 mJ/m^{-2} [36,37]. As a result, at an intermediate strain level, dislocation cell (DC) refinement is the predominant deformation mechanism, while at higher strains, deformation twins are more common [38]. Deformation twinning (DT) was seen in the SLMed AISI 316L at the early stage (3% strain level) of plastic deformation [19,39] and is thought to be a key factor in the material's exceptional ductility (Figure 5.7). The lowered SFE brought on by what may have been a high nitrogen concentration has been linked to this active DT. Furthermore, Liu and co-workers [40] discovered that the cellular substructure exhibits good YS and ductility due to its ability to obstruct moving dislocations but not pin them. Using data from a synchrotron in situ tensile test, Zhang and co-workers [41] also concluded that the high ductility is caused by the low dislocation barrier strength. Additionally, the strain-induced phase transformation behaviour is significantly influenced by the SLMed microstructure.

These outstanding studies have shown that SLMed AISI 316L deformation mechanisms differ considerably from those of conventionally treated materials. Deformation faulting (DF) and DC refinement, the other two primary deformation processes that have been well identified and have an important part in traditionally treated FCC materials, have received minimal consideration in SLMed AISI 316L. The investigation of the SLMed AISI 316L is further constrained by the deficient clarity around these dislocations governed micro-mechanisms. At low strains, DF and DC refinement are the dominant deformation processes in the SLMed AISI 316L. DF and

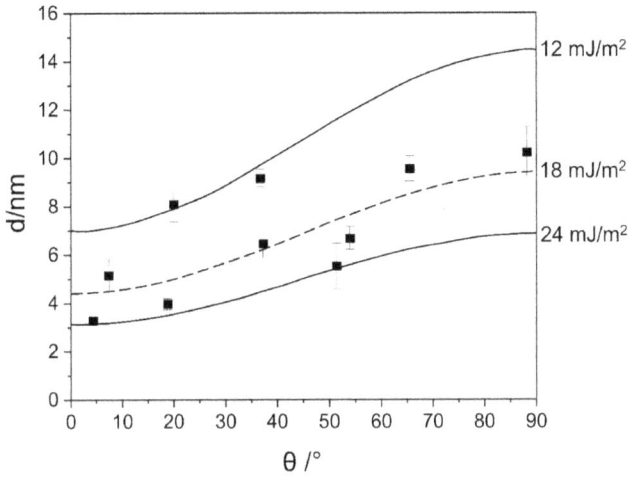

Figure 5.7 Dissociation width of partial dislocations in SLMed AISI 316L [42].

twinning become more pronounced as the strain increases. DF is common in conventional alloys when the SFE is lower, but DC refinement is always present in high-SFE materials. Commencing from this vantage point, it is implausible that DF and DC refining occur concurrently during the tensile behaviour of the SLMed AISI 316L. We conclude that the cellular substructure, the main distinguishing characteristic of SLMed AISI 316L over its conventional equivalent, was responsible for these unanticipated deformation mechanisms.

As a result, in the current section, we have discussed a thorough TEM examination to analyse the deformation substructures and their development under uniaxial tensile stress. These findings demonstrate that the plastic deformation of the SLMed AISI 316L is characterized by DF, DT, and DC refinement. With a focus on how much each process depends on the cellular substructure, DF and DC refinement mechanisms were identified. Based on the observed deformation processes, the strain-hardening behaviour and high ductility of SLMed AISI 316L were addressed.

5.4.1 Deformation faulting

SFE is commonly acknowledged as a primary cause of DF. Furthermore, it was claimed that the SFE of SLMed AISI 316L is lower than that of conventionally treated steel and that interstitial atoms such as N and C are the inherent causes for this. However, He et al. [42] carried out an investigation using atom probe tomography (APT) to measure the interstitial atoms in the interiors of the cells. And they reported that the APT results showed no obvious segregation or clustering, and the compositions of N and C are comparable to those of the conventional AISI 316L. In addition, weak beam dark field (WBDF) analysis can be done to measure the SFE of SLMed AISI 316L [39]. The SFE shown in Figure 5.7 is calculated to be 18 ±6 mJ/m², which is quite compatible with the value discovered using neutron diffraction by Woo and co-workers [43]. The SLMed sample's data points are visibly scattered (Figure 5.7). The reason could be the existence of internal residual stress for this dispersion [44]. Both WBDF and neutron diffraction study results are within the earlier stated value of SFE ranges of AISI 316L that has undergone standard processing [36,37]. Consequently, we can infer that the SFE is not always the rationale behind what caused the DF of SLMed AISI 316L. So, it seems logical to think that the DF is greatly aided by the dislocations created during SLM. He et al. studied the dislocation structures of the SLMed AISI 316L to confirm this presumption. They showed the features of the dislocations in Figure 5.8a. SFs of smaller widths were seen in both cellular edges and interiors, except for cellular boundaries with extensive dislocations. Voisin and co-workers [45] have also confirmed the occurrence of SFs in the as-built SLMed AISI 316L. Experimental and computational research on the first dislocations' creation mechanism has been done [46,47]. It is commonly acknowledged that the cyclic thermal distortions that occur during AM are the cause of dislocations, and it is proposed that the presence of SFs might be owing to small local SFEs brought on by segregation [45] and cyclic deformation as reported for fatigued AISI 316L based on these formation mechanisms [48]. He and co-workers [42] carried out minor distortion of the specimen and analysed the resulting dislocation structures to explain the influence of these detected SFs on the deformation process. The sample dislocation structures of the 1%-deformed AISI 316L are shown in Figure 5.8b. To evaluate the length and breadth of the SFs, they employed similarly imaging circumstances for specimens stretched to distinct levels. The breadth of the SFs in the 1%-deformed sample is substantially bigger than in the as-built sample, as seen in Figure 5.8b. Lee and co-workers [49,50] have shown that broad SFs originate in the original SFs or the very adjacent region. Wide SFs tend to develop at the same or close slip planes, even if they may be isolated by cellular borders, as indicated by the arrows in Figure 5.8b, which closely resemble the distribution of SFs in the as-built specimen (Figure 5.8a). As a result, it can be inferred from these observations that the first SFs serve as nuclei for DF. This idea can also be supported by the development of stacking fault ribbons (SFRs). In the 15% distorted sample, Figure 5.8c provides a typical illustration of how the SFs overlapped and created an initial SFR. Original SFs are present in every cellular interior (Figure 5.8a); thus, when broad SFs arise in the same or nearby slip planes, a long SFR that penetrates the cellular borders emerges (Figure 5.8b). Furthermore, as shown in Figure 5.8c, Liu et al. [40] discovered that the SFs may breach the cellular

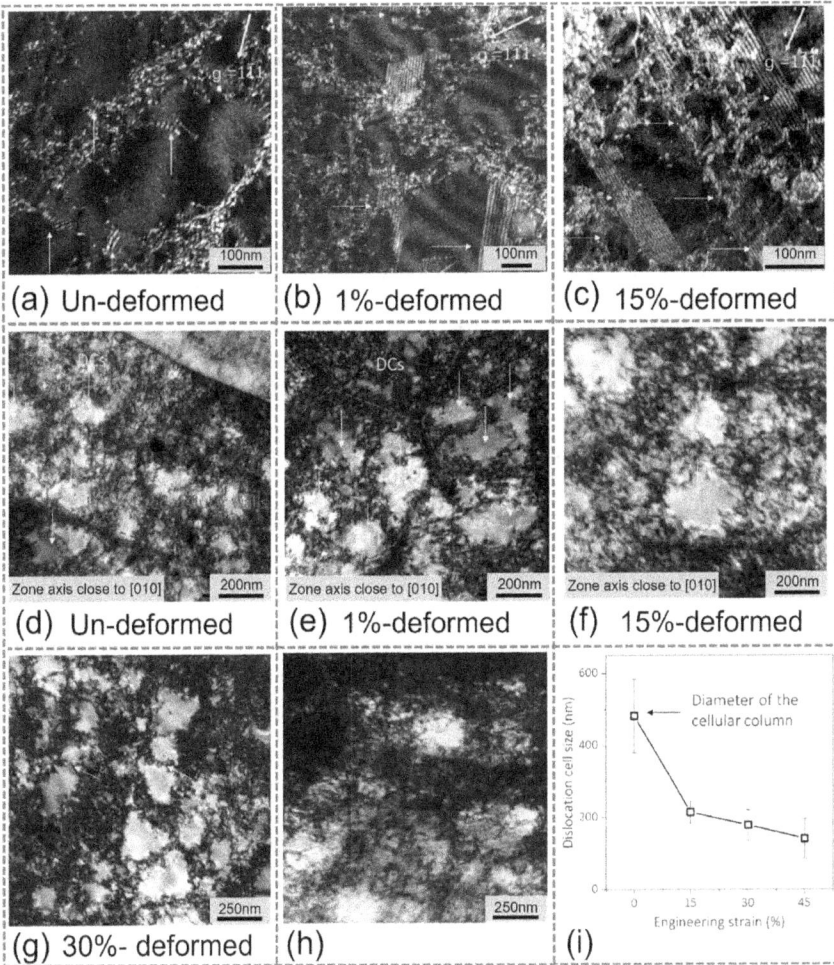

Figure 5.8 (a–c) WBDF images and (d–f) two-beam BF images of un-deformed and various deformed SLMed AISI 316L specimens. (h) Two-beam BF images of the fractured specimen, and (i) the development of dislocation cell size as a function of strain [42] with permission from Elsevier.

borders under the application of strain, which makes it considerably simpler for the long SFR to occur throughout deformation. Further stretching causes the long SFRs to appear once enough broad SFs have adequately overlapped one another.

We get a deeper knowledge of the dislocation dynamics in SLMed AISI 316L by examining the development of pre-existing SFs and their associations with cellular borders. 3D discrete dislocation dynamics simulations in the study of Voisin and co-workers [45] showed that dislocations localized on the cellular boundaries without DF in the early period of deformation. Although it appears from our experimental findings that dislocations favour dissociating into partial dislocations with broad SFs rather than transferring themselves straight to cellular boundaries. Hence, discrete dislocation dynamics models that consider the primary dislocation structures and their responses are recommended in order to explain the mechanical response of SLMed AISI 316L numerically.

5.4.2 Dislocation cell refinement

i. Deformation-induced dislocation cell vs. pre-existing cellular substructure

We first go through the distinction between the pre-existing cellular substructure and the deformation-induced DCs since the as-built sample has pre-existing cellular boundaries with a high DD. The pre-existing cellular substructure is [001] oriented, which means it is columnar in three dimensions, as previous research [45,46] has supported this characteristic. The deformation-induced DCs, on the other hand, are spherical, as they displayed the identical form as imaged from various crystal orientations, as illustrated by the 15%-deformed specimen in Figure 5.8d–f. The pre-existing cellular boundaries are indicated by the red dotted lines in Figure 5.8d. Yellow arrows indicate the location of equiaxed DCs with a size of around 210 nm that were found inside the cellular substructures. By analysing the sample from the [001] zone axis, the cellular substructures in Figure 5.8e were exhibited in an end-on orientation, leading to an equiaxed form (red dotted lines) once more. Despite a 90° tilt in the imaging direction, spherical DCs measuring 210 nm was found inside the cellular substructures (yellow arrows). When seen from the [011] zone axis, the shape of the cellular substructure will be stretched, although the deformation-induced DCs will remain spherical and identical in radius. The shape and size of the DCs, which are independent of imaging direction, revealed that they are produced by tensile deformation and are spherical in 3D.

ii. Dislocation cell refinement upon straining

As shown in Figures 5.9g and h, the DCs are also identified in specimens with a 30% deformation and fracture (~45%). At these strain levels, the distinction between the DC walls and cellular boundaries is diminishing, and the equiaxed DC walls thicken in comparison to those at low stresses. Quantitative evaluation of the DCs' area-equivalent diameter under various stresses has been reported, as shown in Figure 5.8g. The size of DCs [45,46] reduces approximately from 480 to 210 nm after the specimen is strained to 15%, but only slowly from 210 to 140 nm as the strain rises from 15% to 45%. This obvious DC refinement implies a large reduction in the mean free path for dislocation slip throughout the deformation, showing that just considering the interspacing of deformation twins/bands is insufficient for a fundamental assessment of SLMed alloys [51]. The DC size development in Figure 5.8i will offer microstructural factors for the crystal plasticity modelling technique to predict the mechanical response of SLMed AISI 316L.

As stated by the low-energy-dislocation-structure hypothesis [52], DCs emerge throughout deformation in medium-to-high-SFE materials as a result of the intricate communication of dislocations with distinct Burgers vectors. The SFE of SLMed (Figure 5.7) is quite close to that of standard AISI 316L. When the DD is fair enough, SLMed AISI 316L creates DCs on its own. According to prior research [41], the initial DD of the SLMed AISI 316L is around $\sim 2\text{--}4 \times 10^{14}$ m^{-2}, whereas according to the scaling law for AISI 316L, this correlates to a DC size of about 370 nm. Since the primary dendrite arm spacing (PDAS), or the interspacing of segregations, is roughly represented by this density-dependent DC size in the SLMed AISI 316L, dislocations naturally interact and overlap with the micro-segregations produced by rapid solidification to form the [001]-oriented cellular substructures. As the plastic deformation starts, the DD rises according to the equation $\sigma_f = \sigma_{0.2} + M\alpha Gb\rho^{1/2}$ where σ_f is the flow stress, $\sigma_{0.2}$ is the yield stress, M is the polycrystalline Taylor factor, α is a constant, G is the shear modulus, and b is the magnitude of the Burgers vector, which results in a reduction in DC size. When DD hits a certain number, DCs become considerably smaller than PDAS, micro-segregations are no longer the most energetically advantageous region for dislocations to form, and new DCs appear inside the cellular interiors. In theory, when the flow stress increases, the DCs diminish until they reach a critical size. The critical size of DCs in conventionally treated AISI 316L is stated to be 300–400 nm subsequent to application of tensile stress to failure at ambient temperature. He and co-workers discovered that the size of DCs in SLMed AISI 316L is barely half that of standard 316L SS. The cellular substructure of the SLMed AISI 316L should account for this. On one side, DF and twinning are more apparent, resulting in additional surfaces for dislocation storage; on the other, the cellular substructures operate as effective impediments for dislocation trapping, hence increasing DD. This finding is furthermore in line among the research of Zhang and co-workers [41], who found that the DD of the SLMed AISI 316L after tensile testing is two times that of the standard 316L SS. The refining of DCs in SLMed AISI 316L was accomplished by either the dissociation of existing dislocation cell walls (DCWs) or the formation of new DCWs inside existing DCs. The dissociation of existing DCWs at various phases was observed, as indicated by the red arrow in Figure 5.8g. Eventually, additional DCWs appeared in the same area's dislocation-free region; see yellow arrows. This DC refining method is consistent with reported findings in traditionally treated FCC alloys. Though, the pre-existing DCWs do not appear to dissolve under stress since their thickness rises with strain. The trapping action of separated Mo and Cr explains this.

5.4.3 Strain-hardening mechanism and ductility

Based on the microstructure development demonstrated earlier, we illustrated the strain-hardening processes of the SLMed AISI 316L in Figure 5.9. We analysed the strain-hardening mechanism of traditionally treated AISI 316L from the available reported studies to show up the distinctive feature of the SLMed AISI 316L. DF and DC refinement, as opposed to DT, as previously noted [53], are the main strain-hardening processes at stresses of less than 15% in the SLMed 316L SS (Figure 5.9a1 and a2). DT becomes apparent when the strain exceeds 15% while the SFRs and DCs are improved (Figure 5.9a3). Secondary DTs and SFs can be seen at the late stages of deformation (Figure 5.9a4). This strain-hardening technique clearly differs from the standard AISI 316L. According to Figure 5.9b, when the strain is less than 30%, the standard AISI 316L stainless steel is strain hardened by DC refining, and DT emerges as the strain grows. The strain-hardening behaviours and exceptional ductility may be easily understood in light of the strain-hardening processes outlined above. It was observed that the

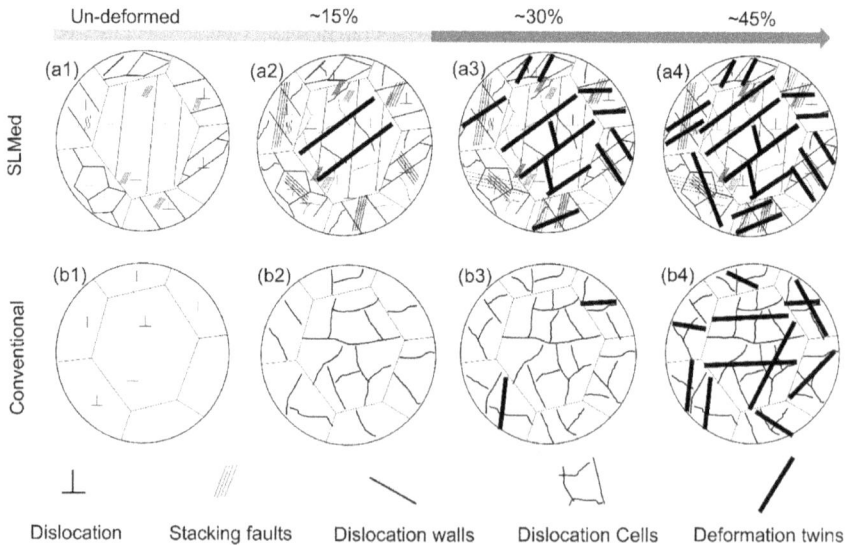

Figure 5.9 Schematic illustrations of the substructure development of 316L SS treated by various techniques with strain level; (a) as-built condition and (b) forged and annealed from literature [42].

initial SLMed AISI 316L specimen had a high amount of dislocation, which led to a lower strain-hardening rate (1.2 GPa) as compared to the wrought AISI 316L specimen (2 GPa). This contrasts with the conventional 3-stage strain-hardening behaviours in conventional FCC materials. These dislocations, however, are distinct from DCs, which significantly reduce the ductility of cold-rolled AISI 316L [54]. Instead, the majority of dislocations were snared by the distinct cellular borders, creating extensive free routes in the [001] direction (Figure 5.9).

While this is going on, several minor SFs are present in the initial microstructure (Figure 5.8a), which makes it simple to develop DF [55,56]. The effectiveness of the SFs in enhancing the strain hardenability of FCC alloys has long been acknowledged [57,58]. Wei and co-workers [58], in particular, recently confirmed the DF driven hardening characteristics of a CoCrNiW alloy, which is consistent with the current reported mechanism. The formation of sessile stair-rod dislocation at the junction of SFs in two crossed slip planes, as indicated by the arrow in Figure 5.8b, can cause Lomer–Cottrell locking. Additionally, the cellular borders might obstruct the partial dislocations that create the SFs (Figure 5.8b), which naturally causes strain hardening, especially in the early stages of deformation [40]. In addition to DF, the SLMed AISI 316L also exhibited the conventional strain-hardening process, or DC refinement (Figure 5.8i). The strain hardenability of SLMed AISI 316L is improved by the DCs' smaller size compared to ordinary 316L SS. As a result, even if the SLMed AISI 316L strain-hardening rate is modest and the initial DD is as high as magnitude 10^{14}, the strain-hardening process nevertheless proceeds gradually.

When the strain reaches 15%, DT becomes apparent, then dominates at elevated strain levels, while the DC refinement turns weak at these strains (Figure 5.8i). Through a dynamic Hall-Petch process, both the improvement of SFRs and DTs result in strain hardening at this point. At 20% strain levels, the dislocation storage rate changes, according to the viewpoint reached by Zhang and co-workers [41] by in situ high-energy X-ray investigation, and this alteration of the strain-hardening process found through microstructure analysis is extremely compatible with that finding. As a result, we draw the conclusion that, for the SLMed 316L SS, DF and DC refinement predominate the behaviour of strain hardening at stresses lower than 15%. The constant strain-hardening rate at high stresses are caused by DT and faulting. The SLMed AISI 316L exhibits exceptional ductility because of the prominent DF and DT, along with the DC refinement.

5.5 CURRENT CHALLENGES AND FUTURE SCOPE

In this chapter, we have systematically explained the influence of HTs on microstructure evolution and mechanical properties. We also stated various deformation mechanisms governing the deformation in SLMed AISI 316L.

However, there is wide scope for further studies. Current shortcomings in existing literature are given below:

- Apart from the diverse literature present on various types of HTs, there is still no systematic study available that considers various types of cooling processes at a wide range of temperatures. Like considering various temperatures followed by different types of cooling methods.
- Deformation mechanisms are available only for as-built or non-heat-treated SLMed samples. There is a wide scope for carrying out studies on heat-treated SLMed AISI 316L samples and discussing various deformation mechanisms occurring inside the SLMed samples.
- Also influence of HTs on crystallographic texture is still missing for SLMed AISI 316L samples.

Thus, all these present challenges can be taken as future scope for further future studies.

5.6 SUMMARY

SLM, one of the most popular AM approaches, has great promise for producing complex-shaped components that are challenging to fabricate using traditional methods. AISI 316L is a commonly used ASS with strong corrosion resistance and exceptional mechanical characteristics, making it a great choice for many industries, including the biomedical, aerospace, defence, oil and gas, and petrochemical sectors. The most recent studies on various HTs and their effect on the microstructures, mechanical properties, and deformation mechanisms of AISI 316L produced by SLM are reviewed in this book chapter. It has been discovered that most of the earlier studies sought to identify the effect of HT leading to modifications in microstructure, and measures have been taken to enhance the mechanical properties of AISI 316L components made using the SLM method. It should be noted that despite these efforts, there are still a number of issues that need to be considered and resolved in follow-up studies, as mentioned in the previous section. Following is the summary of research carried out on AISI 316L by SLM considering various HTs and mechanical properties to date:

- SLMed AISI 316L as a built specimen showed melting pool boundaries, columnar grain structure, as well as cellular grain structure. However, application of high temperature HTs led to the complete disappearance of the melting pool and associated substructures as well as its boundaries, and the entire heat-treated sample consisted of newly formed recrystallized grains with coarse morphology.

- Application of various types of HTs leading to an increase in temperature resulted in a decrease in YS, UTS, and micro-hardness. Drop in dislocation density inside the cell structure is what causes this reduction. In some special cases, an increase in temperature might lead to an increase in YS, UTS, and micro-hardness because of the existence of newly generated silicate precipitates.
- DT, DF, and DC refinement are the prominent deformation mechanisms that all contribute to the superior ductility of SLMed 316 SS by maintaining a constant strain-hardening rate throughout the entire tensile deformation.

ABBREVIATIONS

3D	Three dimensional
AM	Additive manufacturing
ASS	Austenitic stainless steels
BP	Build plane
DC	Dislocation cell
DCW	Dislocation cell wall
DD	Dislocation density
DF	Deformation faulting
DT	Deformation twinning
HAGB	High-angle grain boundary
MP	Melt pool
MPB	Melt-pool boundary
SFE	Stacking fault energy
SFR	Stacking fault ribbon
SLM	Selective laser Melting
SLMed	Selective laser Melted
TB	Twin boundary
UTS	Ultimate tensile strength
YS	Yield strength

REFERENCES

[1] Li, N., Huang, S., Zhang, G., Qin, R., Liu, W., Xiong, H., Shi, G. & Blackburn, J. (2019). Progress in additive manufacturing on new materials: a review. *Journal of Materials Science and Technology*, 35(2), 242–269. https://doi.org/10.1016/j.jmst.2018.09.002.

[2] Haghdadi, N., Laleh, M., Moyle, M. & Primig, S. (2021). Additive manufacturing of steels: a review of achievements and challenges. *Journal of Materials Science*, 56(1), 64–107. https://doi.org/10.1007/s10853-020-05109-0.

[3] Bremen, S., Meiners, W. & Diatlov, A. (2012). Selective laser melting. A manufacturing technology for the future? *Laser Technik Journal, 9*, 33–38. https://doi.org/10.1002/latj.201290018.

[4] Frazier, W. E. (2014). Metal additive manufacturing: a review. *Journal of Materials Engineering and Performance, 23*(6), 1917–1928. https://doi.org/10.1007/s11665-014-0958-z.

[5] Zhang, B., Li, Y. & Bai, Q. (2017). Defect formation mechanisms in selective laser melting: a review. *Chinese Journal of Mechanical Engineering (English Edition), 30*(3), 515–527. https://doi.org/10.1007/s10033-017-0121-5.

[6] Zhai, W., Zhou, W., Zhu, Z. & Nai, S. M. L. (2022). Selective laser melting of 304L and 316L stainless steels: a comparative study of microstructures and mechanical properties. *Steel Research International, 93*(7), 2100664. https://doi.org/10.1002/srin.202100664.

[7] Kruth, J. P., Froyen, L., Van Vaerenbergh, J., Mercelis, P., Rombouts, M. & Lauwers, B. (2004). Selective laser melting of iron-based powder. *Journal of Materials Processing Technology, 149*(1–3), 616–622. https://doi.org/10.1016/j.jmatprotec.2003.11.051.

[8] Bajaj, P., Hariharan, A., Kini, A., Kürnsteiner, P., Raabe, D. & Jägle, E. A. (2020). Steels in additive manufacturing: a review of their microstructure and properties. *Materials Science and Engineering: A, 772*, 138633. https://doi.org/10.1016/j.msea.2019.138633.

[9] Marshall, P. (1984). *Austenitic Stainless Steels: Microstructure and Mechanical Properties.* Springer Science & Business Media, Dordrecht, Netherland Springer Netherlands.

[10] Singh, S., Ramakrishna, S. & Singh, R. (2017). Material issues in additive manufacturing: a review. *Journal of Manufacturing Processes, 25*, 185–200. https://doi.org/10.1016/j.jmapro.2016.11.006.

[11] Ngo, T. D., Kashani, A., Imbalzano, G., Nguyen, K. T. Q. & Hui, D. (2018). Additive manufacturing (3D printing): a review of materials, methods, applications and challenges. *Composites Part B: Engineering, 143*(December 2017), 172–196. https://doi.org/10.1016/j.compositesb.2018.02.012.

[12] Fayazfar, H., Salarian, M., Rogalsky, A., Sarker, D., Russo, P., Paserin, V. & Toyserkani, E. (2018). A critical review of powder-based additive manufacturing of ferrous alloys: process parameters, microstructure and mechanical properties. *Materials & Design, 144*, 98–128. https://doi.org/10.1016/j.matdes.2018.02.018

[13] Dandekar, T. R., Khatirkar, R. K., Mahadule, D., Chavhan, J. & Kumar, D. (2022). Aging phenomenon in low molybdenum Fe-21Cr-5Mn-1.5Ni alloy: microstructure evolution, texture development and corrosion behavior. *Materials Today Communications, 33*(November), 104913.

[14] Ramdas, T. & Aman, D. (2021). Evolution of microstructure and texture in UNS S32750 super duplex stainless steel weldments. *Transactions of the Indian Institute of Metals, 74*(9), 2267–2283. https://doi.org/10.1007/s12666-021-02274-x.

[15] Ramdas, T., Kisni, R., Kumar, A. & Bibhanshu, N. (2022). Unidirectional cold rolling of Fe-21Cr-5Mn-1. 5Ni alloy - microstructure, texture and magnetic properties journal of magnetism and magnetic materials unidirectional cold rolling of Fe-21Cr-5Mn-1. 5Ni alloy - microstructure, texture and magnetic proper. *Journal of Magnetism and Magnetic Materials, 549*(February), 169040. https://doi.org/10.1016/j.jmmm.2022.169040.

[16] Godec, M., Zaefferer, S., Podgornik, B., Šinko, M. & Tchernychova, E. (2020). Quantitative multiscale correlative microstructure analysis of additive manufacturing of stainless steel 316L processed by selective laser melting. *Materials Characterization, 160*, 110074. https://doi.org/10.1016/j.matchar.2019.110074.

[17] Yang, J., Han, J., Yu, H., Yin, J., Gao, M., Wang, Z. & Zeng, X. (2016). Role of molten pool mode on formability, microstructure and mechanical properties of selective laser melted Ti-6Al-4V alloy. *Materials and Design, 110*, 558–570. https://doi.org/10.1016/j.matdes.2016.08.036.

[18] Wang, Y. M., Voisin, T., McKeown, J. T., Ye, J., Calta, N. P., Li, Z., Zeng, Z., Zhang, Y., Chen, W., Roehling, T. T., Ott, R. T., Santala, M. K., Depond, P. J., Matthews, M. J., Hamza, A. V. & Zhu, T. (2018). Additively manufactured hierarchical stainless steels with high strength and ductility. *Nature Materials, 17*(1), 63–70. https://doi.org/10.1038/NMAT5021.

[19] Wang, D., Song, C., Yang, Y. & Bai, Y. (2016). Investigation of crystal growth mechanism during selective laser melting and mechanical property characterization of 316L stainless steel parts. *Materials and Design, 100*, 291–299. https://doi.org/10.1016/j.matdes.2016.03.111.

[20] Özer, G. & Kisasöz, A. (2022). The role of heat treatments on wear behaviour of 316L stainless steel produced by additive manufacturing. *Materials Letters, 327*, 133014. https://doi.org/10.1016/j.matlet.2022.133014.

[21] Tascioglu, E., Karabulut, Y. & Kaynak, Y. (2020). Correction to: influence of heat treatment temperature on the microstructural, mechanical, and wear behavior of 316L stainless steel fabricated by laser powder bed additive manufacturing. *International Journal of Advanced Manufacturing Technology, 107*(5–6), 1957. https://doi.org/10.1007/s00170-020-05115-1.

[22] Tolosa, I., Garciandía, F., Zubiri, F., Zapirain, F. & Esnaola, A. (2010). Study of mechanical properties of AISI 316 stainless steel processed by "selective laser melting", following different manufacturing strategies. *International Journal of Advanced Manufacturing Technology, 51*(5–8), 639–647. https://doi.org/10.1007/s00170-010-2631-5.

[23] Hitzler, L., Hirsch, J., Heine, B., Merkel, M., Hall, W. & Öchsner, A. (2017). On the anisotropic mechanical properties of selective laser-melted stainless steel. *Materials, 10*(10), 1136. https://doi.org/10.3390/ma1010113.

[24] Meier, H. & Haberland, C. (2008). Experimentelle Untersuchungen zum Laserstrahlgenerieren Metallischer Bauteile. *Materialwissenschaft Und Werkstofftechnik, 39*(9), 665–670. https://doi.org/10.1002/mawe.200800327.

[25] Mertens, A., Reginster, S., Paydas, H., Contrepois, Q., Dormal, T., Lemaire, O. & Lecomte-Beckers, J. (2014). Mechanical properties of alloy Ti-6Al-4V and of stainless steel 316L processed by selective laser melting: influence of out-of-equilibrium microstructures. *Powder Metallurgy, 57*(3), 184–189. https://doi.org/10.1179/1743290114Y.0000000092.

[26] Röttger, A., Geenen, K., Windmann, M., Binner, F. & Theisen, W. (2016). Comparison of microstructure and mechanical properties of 316L austenitic steel processed by selective laser melting with hot-isostatic pressed and cast material. *Materials Science and Engineering A, 678*(April), 365–376. https://doi.org/10.1016/j.msea.2016.10.012.

[27] Carlton, H. D., Haboub, A., Gallegos, G. F., Parkinson, D. Y. & MacDowell, A. A. (2016). Damage evolution and failure mechanisms in additively manufactured stainless steel. *Materials Science and Engineering A, 651*, 406–414. https://doi.org/10.1016/j.msea.2015.10.073.

[28] Kong, D., Ni, X., Dong, C., Zhang, L., Man, C., Yao, J., Xiao, K. & Li, X. (2018). Heat treatment effect on the microstructure and corrosion behavior of 316L stainless steel fabricated by selective laser melting for proton exchange membrane fuel cells. *Electrochimica Acta*, 276, 293–303. https://doi.org/10.1016/j.electacta.2018.04.188.

[29] Kong, D., Dong, C., Ni, X., Zhang, L., Yao, J., Man, C., Cheng, X., Xiao, K. & Li, X. (2019). Mechanical properties and corrosion behavior of selective laser melted 316L stainless steel after different heat treatment processes. *Journal of Materials Science and Technology*, 35(7), 1499–1507. https://doi.org/10.1016/j.jmst.2019.03.003.

[30] Mower, T. M. & Long, M. J. (2016). Mechanical behavior of additive manufactured, powder-bed laser-fused materials. *Materials Science and Engineering A*, 651, 198–213. https://doi.org/10.1016/j.msea.2015.10.068.

[31] Spierings, A. B., Starr, T. L. & Wegener, K. (2013). Fatigue performance of additive manufactured metallic parts. *Rapid Prototyping Journal*, 19(2), 88–94. https://doi.org/10.1108/13552541311302932.

[32] Chen, N., Ma, G., Zhu, W., Godfrey, A., Shen, Z., Wu, G. & Huang, X. (2019). Enhancement of an additive-manufactured austenitic stainless steel by post-manufacture heat-treatment. *Materials Science and Engineering A*, 759(January), 65–69. https://doi.org/10.1016/j.msea.2019.04.111.

[33] Hamza, H. M., Deen, K. M., Khaliq, A., Asselin, E. & Haider, W. (2022). Microstructural, corrosion and mechanical properties of additively manufactured alloys: a review. *Critical Reviews in Solid State and Materials Sciences*, 47(1), 46–98. https://doi.org/10.1080/10408436.2021.1886044.

[34] Salman, O. O., Gammer, C., Chaubey, A. K., Eckert, J. & Scudino, S. (2019). Effect of heat treatment on microstructure and mechanical properties of 316L steel synthesized by selective laser melting. *Materials Science and Engineering: A*, 748, 205–212. https://doi.org/10.1016/j.msea.2019.01.110.

[35] Chao, Q., Thomas, S., Birbilis, N., Cizek, P., Hodgson, P. D. & Fabijanic, D. (2021). The effect of post-processing heat treatment on the microstructure, residual stress and mechanical properties of selective laser melted 316L stainless steel. *Materials Science and Engineering: A*, 821, 141611. https://doi.org/10.1016/j.msea.2021.141611.

[36] Lu, J., Hultman, L., Holmström, E., Antonsson, K. H., Grehk, M., Li, W., Vitos, L. & Golpayegani, A. (2016). Stacking fault energies in austenitic stainless steels. *Acta Materialia*, 111, 39–46. https://doi.org/10.1016/j.actamat.2016.03.042.

[37] Molnár, D., Sun, X., Lu, S., Li, W., Engberg, G. & Vitos, L. (2019). Effect of temperature on the stacking fault energy and deformation behaviour in 316L austenitic stainless steel. *Materials Science and Engineering A*, 759(May), 490–497. https://doi.org/10.1016/j.msea.2019.05.079.

[38] Zieliński, W., Abduluyahed, A. A. & Kurzydłowski, K. J. (1998). TEM studies of dislocation substructure in 316 austenitic stainless steel strained after annealing in various environments. *Materials Science and Engineering A*, 249(1–2), 91–96. https://doi.org/10.1016/s0921-5093(98)00578-4.

[39] Li, L., Chen, Z., Kuroiwa, S., Ito, M., Kishida, K., Inui, H. & George, E. P. (2022). Tensile and compressive plastic deformation behavior of medium-entropy Cr-Co-Ni single crystals from cryogenic to elevated temperatures. *International Journal of Plasticity*, 148(October 2021), 103144. https://doi.org/10.1016/j.ijplas.2021.103144.

[40] Liu, L., Ding, Q., Zhong, Y., Zou, J., Wu, J., Chiu, Y. L., Li, J., Zhang, Z., Yu, Q. & Shen, Z. (2018). Dislocation network in additive manufactured steel breaks strength-ductility trade-off. *Materials Today*, 21(4), 354–361. https://doi.org/10.1016/j.mattod.2017.11.004.

[41] Zhang, X., Kenesei, P., Park, J. S., Almer, J. & Li, M. (2021). In situ high-energy X-ray study of deformation mechanisms in additively manufactured 316L stainless steel. *Journal of Nuclear Materials*, 549, 152874. https://doi.org/10.1016/j.jnucmat.2021.152874.

[42] He, F., Wang, C., Han, B., Yeli, G., Lin, X., Wang, Z., Wang, L. & Kai, J. J. (2022). Deformation faulting and dislocation-cell refinement in a selective laser melted 316L stainless steel. *International Journal of Plasticity*, 156(January), 103346. https://doi.org/10.1016/j.ijplas.2022.103346.

[43] Woo, W., Jeong, J. S., Kim, D. K., Lee, C. M., Choi, S. H., Suh, J. Y., Lee, S. Y., Harjo, S. & Kawasaki, T. (2020). Stacking fault energy analyses of additively manufactured stainless steel 316L and CrCoNi medium entropy alloy using in situ neutron diffraction. *Scientific Reports*, 10(1), 2–4. https://doi.org/10.1038/s41598-020-58273-3.

[44] Chen, W., Voisin, T., Zhang, Y., Florien, J. B., Spadaccini, C. M., McDowell, D. L., Zhu, T. & Wang, Y. M. (2019). Microscale residual stresses in additively manufactured stainless steel. *Nature Communications*, 10(1), 1–12. https://doi.org/10.1038/s41467-019-12265-8.

[45] Voisin, T., Forien, J. B., Perron, A., Aubry, S., Bertin, N., Samanta, A., Baker, A. & Wang, Y. M. (2021). New insights on cellular structures strengthening mechanisms and thermal stability of an austenitic stainless steel fabricated by laser powder-bed-fusion. *Acta Materialia*, 203, 116476. https://doi.org/10.1016/j.actamat.2020.11.018.

[46] Bertsch, K. M., Meric de Bellefon, G., Kuehl, B. & Thoma, D. J. (2020). Origin of dislocation structures in an additively manufactured austenitic stainless steel 316L. *Acta Materialia*, 199, 19–33. https://doi.org/10.1016/j.actamat.2020.07.063.

[47] Lindroos, M., Pinomaa, T., Ammar, K., Laukkanen, A., Provatas, N. & Forest, S. (2022). Dislocation density in cellular rapid solidification using phase field modeling and crystal plasticity. *International Journal of Plasticity*, 148(October 2021), 103139. https://doi.org/10.1016/j.ijplas.2021.103139.

[48] Cui, L., Jiang, F., Peng, R. L., Mousavian, R. T., Yang, Z. & Moverare, J. (2022). Dependence of microstructures on fatigue performance of polycrystals: a comparative study of conventional and additively manufactured 316L stainless steel. *International Journal of Plasticity*, 149(November 2021), 103172. https://doi.org/10.1016/j.ijplas.2021.103172.

[49] Lee, E. H., Byun, T. S., Hunn, J. D., Yoo, M. H., Farrell, K. & Mansur, L. K. (2001). On the origin of deformation microstructures in austenitic stainless steel: part I - Microstructures. *Acta Materialia*, 49(16), 3269–3276. https://doi.org/10.1016/S1359-6454(01)00193-8.

[50] Lee, E. H., Yoo, M. H., Byun, T. S., Hunn, J. D., Farrell, K. & Mansur, L. K. (2001). On the origin of deformation microstructures in austenitic stainless steel: part II-Mechanisms. *Acta Materialia*, 49(16), 3277–3287. https://doi.org/10.1016/S1359-6454(01)00194-X.

[51] Motaman, S. A. H. & Haase, C. (2021). The microstructural effects on the mechanical response of polycrystals: a comparative experimental-numerical study on conventionally and additively manufactured metallic materials. *International Journal of Plasticity*, *140*(July 2020), 102941. https://doi.org/10.1016/j.ijplas.2021.102941.

[52] Kuhlmann-Wilsdorf, D. (1999). The theory of dislocation-based crystal plasticity. *Philosophical Magazine A: Physics of Condensed Matter, Structure, Defects and Mechanical Properties*, *79*(4), 955–1008. https://doi.org/10.1080/01418619908210342.

[53] Pham, M. S., Dovgyy, B. & Hooper, P. A. (2017). Twinning induced plasticity in austenitic stainless steel 316L made by additive manufacturing. *Materials Science and Engineering A*, *704*(April), 102–111. https://doi.org/10.1016/j.msea.2017.07.082.

[54] Yan, F. K., Liu, G. Z., Tao, N. R. & Lu, K. (2012). Strength and ductility of 316L austenitic stainless steel strengthened by nano-scale twin bundles. *Acta Materialia*, *60*(3), 1059–1071. https://doi.org/10.1016/j.actamat.2011.11.009.

[55] Byun, T. S. (2003). On the stress dependence of partial dislocation separation and deformation microstructure in austenitic stainless steels. *Acta Materialia*, *51*(11), 3063–3071. https://doi.org/10.1016/S1359-6454(03)00117-4.

[56] Byun, T. S., Lee, E. H. & Hunn, J. D. (2003). Plastic deformation in 316LN stainless steel - characterization of deformation microstructures. *Journal of Nuclear Materials*, *321*(1), 29–39. https://doi.org/10.1016/S0022-3115(03)00195-8.

[57] Fu, Z., MacDonald, B. E., Zhang, D., Wu, B., Chen, W., Ivanisenko, J., Hahn, H. & Lavernia, E. J. (2018). FCC nanostructured TiFeCoNi alloy with multi-scale grains and enhanced plasticity. *Scripta Materialia*, *143*, 108–112. https://doi.org/10.1016/j.scriptamat.2017.09.023.

[58] Wei, S. & Tasan, C. C. (2020). Deformation faulting in a metastable CoCrNiW complex concentrated alloy: a case of negative intrinsic stacking fault energy? *Acta Materialia*, *200*, 992–1007. https://doi.org/10.1016/j.actamat.2020.09.056.

Chapter 6

Advancements in integrated additive manufacturing for composite materials

Techniques, challenges, case studies, and applications

Praveen Kumar
Sant Longowal Institute of Engineering and Technology

Palanisamy Sivasubramanian
Dilkap Research Institute of Engineering and Management Studies

Pradeepkumar C
Kalasalingam Academy of Research and Education

Nitin Yadav
Sant Longowal Institute of Engineering and Technology

Carlo Santulli
Università degli Studi di Camerino

6.1 INTRODUCTION

Manufacturing is the process of converting raw materials into finished products. It is a complex activity that involves a number of steps, including the design and production of machines, tools, and fixtures (Sefene et al., 2022). The history of manufacturing dates back to prehistoric times, when people began using stone tools to shape their food. This led to the development of agriculture, which required more sophisticated tools and techniques. The industrial revolution began in Britain with the invention of steam engines in the late 18th century (Popov & Fleisher, 2020). These machines were used to pump water out of mines and then powered mills that ground grain into flour or crushed coal into coke for heating homes and factories alike. The use of electricity also led to an increase in automation as well as mechanization, i.e., the use of machines instead of humans, which allowed factories to produce more goods per worker than ever before possible before this time period (19th century).

DOI: 10.1201/9781003406488-6

Additive manufacturing (AM), or 3D printing, is not, in the narrowest sense, a new technology. It has been around for decades, but it is used today to change the way we produce things. The history of AM dates back to the 1980s, when a company called 3D Systems introduced its first product: a machine that could produce plastic parts (Kumar et al., 2022a, 2022b). This was followed by many other similar products, all based on this same concept: using a computer program to create designs in three dimensions (3D) and then printing them out with an automated process. The technology was first used in small-scale applications, like producing replacement parts for cars or jewelry. Since then, it has evolved over time into something much more powerful than anything previously experimented with, so that it is now possible to create entire objects from scratch by simply clicking on an image file. Today, there are many different types of AM machines accessible on market shelves all over the world, and most of them use different technologies such as laser sintering or stereolithography (SLA) (Deshwal et al., 2020).

Some of the advantages of AM over subtractive include, as from Pragana et al. (2021):

> *Greater design freedom*: with AM, there is no requirement to create a negative or inverse of your design to produce it. This allows for much more intricate design and geometries that would be difficult to obtain by means of traditional manufacturing approaches.
> *Faster production times*: AM can be significantly faster than subtractive methods, especially when producing small batch sizes or prototypes.
> *Reduced waste*: AM generates very little waste material, as compared to subtractive processes, which generate large amounts of scrap material that must be disposed of.
> *Lower energy consumption*: AM generally requires less energy than traditional methods, in particular it does not rely on large amounts of mechanical/thermal energy to remove material, as it is instead the case in welding or machining.

The history of hybrid additive manufacturing (HAM) goes back to the early 2000s, when researchers at the Massachusetts Institute of Technology (MIT) first began to explore the potential of combining 3D printing with injection molding (Boros et al., 2019). The idea behind this type of hybrid fabrication is that both additive and subtractive methods are used simultaneously to create a piece, which is then finished using traditional machining processes (Pawlowski et al., 2017). In 2007, MIT researchers published their findings on this new process, which they called "additive-injection molding," or AIM. This method allowed them to combine the advantages of both AM and injection molding into one process, conferring on it the ability to create intricate objects on demand without requiring expensive tooling or large amounts of material. The advantages of HAM include the ability to build parts with high precision and accuracy, as well as the ability to print functional materials such as metal (Ngo et al., 2018). This is important because it allows for the manufacturing of parts that are light and strong, which can be used in applications such as aerospace engineering or automotive manufacturing (Dilberoglu et al., 2021).

Different HAM techniques include combining multiple processes to create parts or products. This can involve using multiple 3D printing technologies together, such as depositing material onto a build platform and then laser melting it. It can also involve combining additive and subtractive processes, such as machining and 3D printing. HAM can offer advantages over traditional manufacturing methods, such as increased flexibility and speed and reduced waste (Zhu et al., 2013).

6.2 HYBRID ADDITIVE MANUFACTURING (HAM) FOR METALS

AM of metals is a process of joining materials to create objects from 3D models (Mellor et al., 2014). It is an emerging technology that has the potential to transform the way we manufacture products. Metal AM is a rapidly growing field with immense potential. In 2015, the global market for metal AM was estimated at $2.9 billion and is expected to grow to $12.8 billion by 2025. The key drivers for this growth are the unique advantages that AM offers over traditional subtractive manufacturing techniques (Jiménez et al., 2021).

6.2.1 Hybrid additive manufacturing (HAM) for metals based on forming

HAM is a grouping of substantial extrusion 3D printing procedures, general manufacturing processes, and some AM types (e.g., powder-bed fusion (PBF), directed energy deposition (DED), and sheet lamination (SL)). The laser-based PBF method is popular for the precision and surface finish achieved with it. DED systems have advantages in terms of speed and larger part capabilities (Popov & Fleisher, 2020). Hybrid processes are grouped into two categories: hybrid 3D printing processes and metal AM processes. Bambach et al. used a hybrid approach of producing and wire arc additive manufacturing (WAAM) to create a propeller from Ti-6Al-4V (Bambach et al., 2020). The pre-shaped, semi-finished part was produced by WAAM, then forged for its final shape. Tests revealed a superior metallurgical attachment on the interface conferred by the dual process and a slightly lower tensile strength compared to forging but greater than casting by 5%. Ambrogio et al.'s research combined AM with single-point incremental creation to enhance the mechanical strength performance of a printed tank upper cover (Ambrogio et al., 2019). SLS fabricated the sheets, while single-point incremental forming addressed the remaining geometry; this increased the variety of shape typology while improving accuracy and avoiding excessive sheets. Liravi and Toyserkani conducted a study to investigate combining jetting and extrusion processes for medical applications (Liravi & Toyserkani, 2018). The objects are printed using electromechanical, magnetic, and pneumatic nozzles. The outcome demonstrated better printing firmness and external quality in objects ranging 8–10 microns root means square of waviness for binder jetting 3–4 microns for the extrusion process at the best settings (Meda, 2021). Additionally, it was shown that microhardness declined as deposition height rose, although this tendency attenuated as more heat was collected.

6.2.2 Hybrid additive manufacturing (HAM) for metals based on machining

HAM for metals is a process that combines traditional machining with 3D printing technology. This allows for the manufacturing of metal parts with increased accuracy and precision while also reducing the overall cost of production. The HAM process commences with the creation of a 3D model of the desired part (Silva et al., 2019). Once the model is complete, it is sent to a Computer Numerical Control (CNC) machine, where it is milled into a negative mold. This mold is then used to create a positive copy of the part using a 3D printer. The final step in the process is to finish the part using traditional machining techniques such as polishing and surface finishing. The benefits of HAM for metals include increased accuracy, reduced production costs, and shorter processing times. This technology is ideal for applications where precise metal parts are required, such as in the aerospace and medical industries (Hasanov et al., 2022).

The FIPT developed the controlled metal buildup (CMB) method, combining a milling machine with laser deposition (Sefene et al., 2022). This method improved sustainability over a freestanding five-axis milling machine by allowing the use of diverse metals such as bronze, steel, and hard alloys, as well as by continuously engaging milling tools to create deep, narrow grooves. Test parts made from stainless steel emerged from this process with more than 95% density and an impressive surface quality (Sidambe, 2014). To demonstrate the application of this process, a tool for injection molding rubber parts was sintered.

In most cases, subtractive manufacturing processes are linked with the milling (and occasionally turning) process, which uses DED technology (Dilberoglu et al., 2021). Generally, a near-net shape is produced through an additive procedure, and then the material is removed afterwards. For complicated features, some components may need intermediate machining between layers, depending on the part's design. Furthermore, in terms of material removal, there are a number of issues that may arise from DED technology, such as residual stresses or thermal distortion caused by high temperatures. As such, it is important to ensure quality control in the additive phase to minimize flaws like pores or cracks. Feeding mechanisms are also essential when dealing with DED technology, as powder size and wire feed lead to a trade-off between deposition rate and porosity (Dornelas et al., 2022). Additionally, parameters such as spindle speed and feed rate must be accounted for during machining as they affect overall production time and cost. The depth of cut is another crucial factor that must be monitored to guarantee dimensional tolerances in the final product. End mills offer certain advantages in this regard due to their high hardness and toughness, but they lack proper cooling mechanisms. Moreover, HAM processes provide adjustable layer thicknesses, which helps reduce production time considerably compared to CNC processes (Chávez et al., 2021).

6.3 INDUSTRIAL APPLICATION FOR HAM

In HAM, the industries produce the tooling product, whereas one self-regulating manufacturing unit is defined for hybrid series, enabling the process flow in the industries according to the product (Sepasgozar et al., 2020), as reported in Figure 6.1. A powder storage area is interconnected with metal-based additive manufacturing (MAM), in which Laser Beam Powder-Bed Fusion (LPBF) is used. The robotic arm follows LPBF to circulate the product from one end to another in the conveyor belt. The heat treatment process with this line production of the effect is transferred to the CNC and Quality Control (QC) (Elhabashy et al., 2019). After that, their finishing step will be conducted and forwarded to packing and dispatch. In this series, the processed product applies to aerospace applications.

Generally, AM technology has not considerably changed recently. The polyjet technique, which uses droplet deposition print heads to deposit photopolymers, is able to give a flair to the latest commercial technology. It is said that significant advances in this technology will make the switch from Rapid Prototyping (RP) to AM justified (Campbell et al., 2012). As was already indicated, this transformation results from several evolutionary

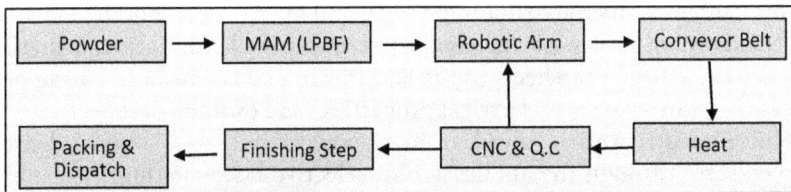

Figure 6.1 Process flow of manufacturing the tooling product.

advancements in the materials and processes and the technology's lower cost, making it more accessible to a broad-range user base. However, there has also been a shift in consumers' perspectives that has made them more open to new applications or even perceive them as a more helpful substitute for current ones. Because technology made it possible for new items to reach the market swiftly and predictably, automotive makers took advantage of it. Short time and money savings might add significant overall savings in constructing a vehicle (Campbell et al., 2012). One way is to use AM as a component of a bridge tooling process to make components available before the tool's total manufacture is complete. Because this is the most economical method, the high-end, lower-volume auto manufacturers are even choosing AM as their preferred production technique for their specific parts. Of course, these competitive advantages would be gone if all automakers used the same technologies as they do now. However, AM is currently a crucial technique for creating automotive products, where, in the last decade, it has appeared as a cutting-edge manufacturing technique with distinct benefits over conventional manufacturing techniques, though bearing still restricted industrial acceptance (Sun et al., 2021).

Several significant disadvantages of AM can be overcome by combining AM with other supporting techniques (Abdulhameed et al., 2019). As a result, HAM has recently received interest from researchers and the modern manufacturing industry, which is generally illustrated in Figure 6.2, and is projected to have further developments in the future for its flexibility (Dilberoglu et al., 2021).

A hybrid-based multitasking machine tool built by adding laser source-based metal deposition technology is already fabricated, installed, and integrated with turning and milling capabilities to satisfy these criteria, opening up new possibilities for the machine tools (Yamazaki, 2016). The theory behind the hybrid multitasking platform provides a further evolution of Done-in-One processes, allowing for the rapid creation of the entire shape through the high accuracy finish in machining operations after producing near-net shape components using AM. It is significantly more likely adapted to smaller lots of manufacturing of higher hardness materials, such as alloys used in aerospace and in the energy industry, the manufacturing of tools and their components with higher precision, and the manufacturing of special alloy products, frequently used for the creation of medical devices (Merklein et al., 2016). The HAM 3D printing technique was used to create functional gradient materials by combining microextrusion, pico-jet, and fiber laser technologies with a multi-axis robot to deposit the conductive traces made for various materials (McKenzie, 2018). The effects of the laser beam and the furnace curing processes on grain shape and electrical properties were examined.

The microstructure evolved from a powder-like to a highly densely packed state through the application of selective laser melting by combining the different effects, especially for Nickel-based superalloys (Heeling et al., 2017). Compared to furnace curing, which took more than 2 hours,

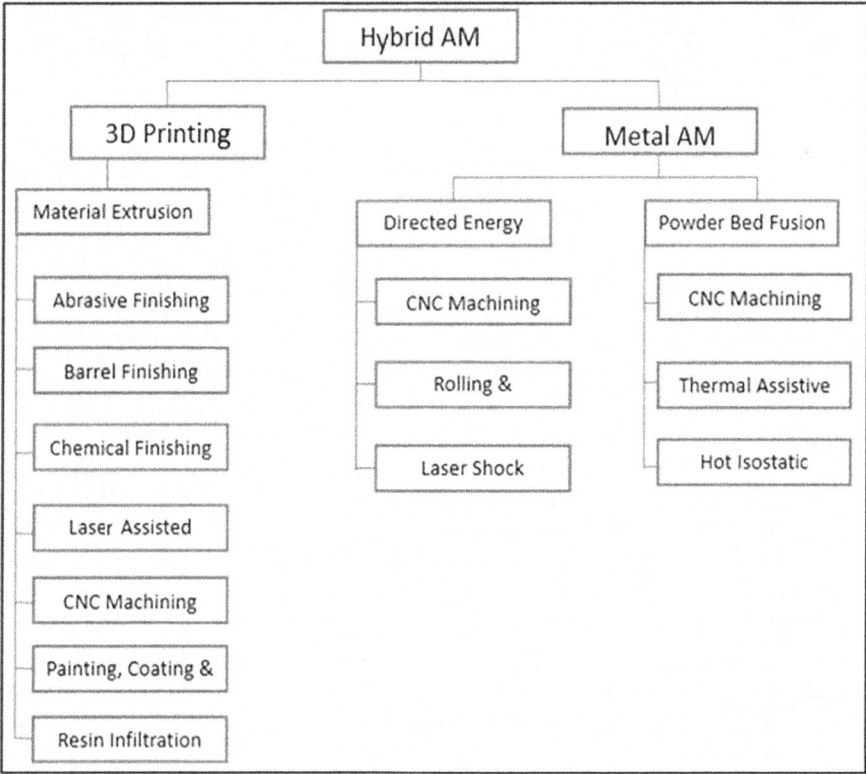

Figure 6.2 Classification of HAM processes.

the laser power curing mechanism reduced resistance with a quicker curing time (1 minute). The ideal sintering parameters were obtained for four laser passes with operation at 18 W with equivalent resistivity to more significant materials (Parupelli & Desai, 2020).

Some vital contemporary developments in HAM for steel-based materials have also been explored (Popov & Fleisher, 2020).

a. Steel-based materials using PBF and DED methods are evolving quickly.

b. Several attempts have reportedly been made to create hybrid and composite materials based on steel using PBF methods.

c. The integration of DED techniques with the conventional CNC process is more flexible. This combination is advantageous for restoring damaged parts, applying protective coatings, and carrying out hybrid production.

d. Laser source-based PBF enables HAM, in which the more specific shape substrate part is created using conventional manufacturing techniques and the complex shape component can be printed directly.

A global background and fundamental principles of AM and the hybrid 3D printing technology that emerged over the past 10 years are offered by Daminabo et al. (2020). The expansion of the definition of fusion manufacturing was evaluated to include the use of materials that are not primarily processed from raw materials in various forms, such as powders, pellets, tubes, rods, sheets, and ingots. A novel hybrid production process based on the systematic, controlled application of the process mechanism to additively deposited materials and the controlled application of AM to primarily processed raw materials that were previously produced through the traditional conventional manufacturing processes is made possible by the newly proposed hybridization (Pragana et al., 2021). Overall, the study demonstrates that incorporating a MAM process with a traditional manufacturing process achieves the essential double priorities of expanding the applicability domain and removing AM drawbacks, such as lower productivity, more metallurgical flaws, poor surface finish, and a lack of dimensional accuracy, and fostering more flexibility in new applications over traditional manufacturing processes or routes.

Aircraft: AM techniques are perfect for this industry because aerospace components must be produced in small batches with complex geometries essential for airflow and heat dissipation functions (Singamneni et al., 2019). Additionally, on-demand and on-site manufacturing of parts for space station maintenance or repair. Besides, AM can create components with low weight-to-strength ratios, which are critical for space shuttles and airplanes. Given the high cost of the materials used in the aircraft sector and the fact that AM methods are known for producing less waste, AM has grown in popularity among aerospace industry producers (Mehrpouya et al., 2019).

Medical: Making medical implants was one of the earliest indications that AM was entering the field of medicine. Medical implants must be patient-specific in addition to having sophisticated designs. AM is cost-efficient for making small lots of components, which is more prevalent in the aerospace and medical sectors than traditional processes (Mehrpouya et al., 2019). The price and length of surgery are also reduced by producing implants unique to each patient. Ti6Al4V hip stems with functionally graduated porosity properties were created by laser-engineered net shaping (LENS).

Automobile: A part of the automotive industry also requires complexity and a low weight-to-strength ratio. The advantages of AM have allowed it to be used for production of and prototyping of automotive parts (Salifu et al., 2022). For instance, Optomec Laser LENS was employed to cut the cost, time, and material requirements for manufacturing components for Red Bull racing cars, such as drive shaft spiders and suspension mounting brackets (Mehrpouya et al., 2019).

The future of construction and infrastructure is also being created using additive construction, according to Camacho et al. (2018). So, as a result, continuous research into concrete and other construction materials provides the steppingstone for 3D printing technology to be used in civil engineering applications. With a day-by-day textile printing advancement, another AM industry branch has grown in the apparel and jewelry sectors. Rapid design (shorter manufacturing duration) and lower packaging and transportation costs are just two of the more significant advantages for AM in the fashion design sector (Attaran, 2017). The key ingredients used in cuisine are another exceptional material for 3D printing. Using extrusion-based AM processes, a wide range of applications with the required surface roughness and diverse nutritional contents have recently been investigated. Although there are still issues with process productivity, material durability, and edible material serviceability, AM may have a future in food production (Dilberoglu et al., 2017).

6.4 CHALLENGES IN HYBRID ADDITIVE MANUFACTURING

Verification of the mechanical and thermal characteristics of the additive layer manufacturing (ALM) technologies and the current materials was investigated by Petrovic et al. (2011). The behavior of additively manufactured pieces differs from that of parts made traditionally in different directions. When the load is applied in the layer direction instead of the buildup direction, they are "stronger." Limited characterization has been carried out as of yet to measure the difference, for example in Reith et al. (2020), therefore also delaying the development, until characterization is customary and standardized. New material development and characterization for ALM. AM technologies process an extensive range of metals and polymers, including titanium, chromium, cobalt, stainless steel, and others. Other intriguing materials require additional study (biodegradable polymers, copper, magnesium, etc.) (Petrovic et al., 2011). A shift in the research developer's perspectives on atomic layer metallization has made it possible for new "freeform fabrication," which is exempt from production restrictions and needs to be publicized.

Designing and planning processes for automation has also substantial importance; whether a production process is forming, subtractive, or additive, the development of automatic assessment tools is a general trend (Flynn et al., 2016). Several outcome criteria are considered while designing for AF, and they may all be leveraged to automate decision-making in ALM. The arrangement of layers, which affects surface quality and roughness, results in the stair-stepping effect (Petrovic et al., 2011). Different zones of the model will have rougher or smoother surfaces depending on how it is oriented for cutting and fabrication. The amount of material utilized for the support structure should be kept to a minimum because it cannot be recycled.

Achieving an effective and sound connection between the components obtained on off-site printed units, the quality of the surface finishes, and the robustness of 3D printing components in the surrounding situation due to the targeted material properties requirements, such as zero slump, are all issues in addition to the manufacturing encounters of 3D printing (Sepasgozar et al., 2020). Building information modeling (BIM)-based 3DP techniques from the past demonstrate efficient information interchange and model updating. However, it has been noted that integrating BIM with 3D printing is complex, and problems such as copyright and modeling duties are still mostly unresolved. The primary consideration has switched from prototyping to BIM and concrete materials, according to a comparison with a few selected peer publications from 2011 to 2013 and 2018–2020 of the previous decade (Li et al., 2020).

AM has many advantages, such as design flexibility, the capacity to produce intricate geometries, ease of usage, and greater product personalization (Alfaify et al., 2020). However, AM technology needs to be developed for applications where it can be used in real-time. There have been downsides and difficulties that call for further inquiry in addition to technological advancement. The challenges that need to be addressed further include size restriction, anisotropy on mechanical properties, construction of overhanging surfaces, higher prices, deprived production efficiency, low precision, pillowing, gaps, warping, stringing, gaps between the top layers, underextrusion, misalignment on the coating, excuse extrusion, elephant foot, mass manufacturing, and material usage restrictions (Abdulhameed et al., 2019).

Since its amalgamation and development more than 25 years ago, quick prototyping, rapid tooling, direct part manufacturing, and restoration have all seen significant changes. AM technologies have shown considerable potential for aircraft applications and can reduce component weight, increase fuel efficiency, and lower costs and turnaround times. However, aerospace applications still have technical difficulties with AM relating to design, material properties, process control, process comprehension, and modeling (Liu et al., 2017). Al alloys have seen less research toward HAM than steels and composite-based materials (Altıparmak et al., 2021). This shortcoming can be attributed to: (i) challenges incorporating AM technology; (ii) the restricted pertinency of aluminum alloys due to metallurgical flaws; and (iii) the lack of accreditation for AM and HAM processes for aluminum aerospace components, which limits the use of AM processes for these parts. Reviewing recently created HAM techniques for Al alloys reveals that they successfully address certain disadvantages by integrating AM processes into traditional production chains. SLM HAM methods, in particular, are efficient in reducing and removing metallurgical flaws to produce dense (flawless) Al parts with improved mechanical qualities (Uçak et al., 2022).

6.4.1 Precise prediction of the properties of the lattice structure

The integration of microscale and macroscale in large-scale hierarchical structural analysis and topology optimization for AM is greatly facilitated by homogenization theory. The asymptotic development of the regulating equations is necessary for homogenization to realize structural analysis at various scales. Homogenization allows for the microscopic-scale simulation of physical fields (such as stress) as well as adequate material characteristics (Matouš et al., 2017). The structural design of periodic microstructures and the multiscale structural analysis of composite materials both frequently employ homogenization. An analogous energy-based homogenization was also suggested to anticipate the useful properties of microstructure and get beyond the challenges of numerical computation and sensitivity derivation. The average stress and strain tensors of the periodic microstructures are the same as the stress and strain tensors of the analogous homogenous medium (Pecullan et al., 1999). Microstructures have the same strain energy density as an analogous homogenous medium. For a 2D orthotropic microstructure, four straightforward load cases are needed to determine the stiffness tensor, but nine are required for 3D examples. Literature has also presented energy-based methods for predicting thermal conductivities and effective coefficients of thermal expansion (Jihong et al., 2021).

6.4.2 Void formation

Creating voids between consecutive layers of material is one of the critical downsides of 3D printing (Blok et al., 2018). Because there is less interfacial bonding between printed layers, the different porosity produced by AM can significantly impact mechanical performance. The method and material used in 3D printing substantially impact how much void is formed. The production of voids is thought to be one of the significant flaws that lead to inferior and anisotropic mechanical qualities in processes that use filaments of materials, such as fused filament fabrication (FFF) or contour crafting (Rajan et al., 2022). After printing, this void development may potentially cause the layers to separate. Increasing the thickness of the filament in a composite that was 3D printed using the FFF method decreased porosity but deteriorated cohesion, which caused a decline in tensile strength and an increase in water absorption. In the additive production of concrete, the thicker concrete layers and longer gaps between them produced better interlayer bonding and reduced void formation. On the other hand, by reducing the height of each layer in the powder-bed printing of an alumina/glass composite, the high porosity of AM can be significantly reduced (Ngo et al., 2018).

6.5 CASE STUDIES

Many case studies are done to show the potential and abilities of the HAM process. To summarize, the HAM process is superior to individual AM or SM processes in that it can: (i) utilize an integrated manufacturing platform to combine additive and subtractive processes into a single place to cut down on manufacturing time and material waste; (ii) enhance surface finish by lowering the possibility of stacked tolerances and by eradicating the staircase mistake with the help of 6-DOF flexibility; (iii) produce freeform surfaces by changing the tool axis direction (TAD) concerning the surface normal vector; (iv) process both additive and subtractive features; and (v) improve the manufacturing system's capacity to avoid collisions through the use of a 6-DOF robotic platform (Li et al., 2018).

6.6 CONCLUSION

The current assessment and study may be used to extrapolate the key common developments in HAM approaches for composites. The goal of this study was to take the readers on a trip that began with an explanation of the background and operating principles of AM and finished with a look at the HAM techniques that have emerged over the past decade. A particular focus was placed on the classification of HAM for the integration of forming processes, namely bulk and sheet forming operations, and procedures to enhance the mechanical characteristics of the additively deposited materials. In conclusion, the HAM process combines the benefits of additive and subtractive manufacturing, integrates HAM into traditional production chains with compensating HAM limitations, and increases HAM processes' industrialization.

REFERENCES

Abdulhameed, O., Al-Ahmari, A., Ameen, W., & Mian, S. H. (2019). Additive manufacturing: challenges, trends, and applications. *Advances in Mechanical Engineering*, 11(2), https://doi.org/10.1177/1687814018822880.

Alfaify, A., Saleh, M., Abdullah, F. M., & Al-Ahmari, A. M. (2020). Design for additive manufacturing: a systematic review. *Sustainability*, 12(19), 7936.

Altıparmak, S. C., Yardley, V. A., Shi, Z., & Lin, J. (2021). Challenges in additive manufacturing of high-strength aluminium alloys and current developments in hybrid additive manufacturing. *International Journal of Lightweight Materials and Manufacture*, 4(2), 246–261.

Ambrogio, G., Gagliardi, F., Muzzupappa, M., & Filice, L. (2019). Additive-incremental forming hybrid manufacturing technique to improve customised part performance. *Journal of Manufacturing Processes*, 37, 386–391.

Attaran, M. (2017). The rise of 3-D printing: the advantages of additive manufacturing over traditional manufacturing. *Business Horizons*, 60(5), 677–688.

Bambach, M., Sizova, I., Sydow, B., Hemes, S., & Meiners, F. (2020). Hybrid manufacturing of components from Ti-6Al-4V by metal forming and wire-arc additive manufacturing. *Journal of Materials Processing Technology*, 282, 116689.

Blok, L. G., Longana, M. L., Yu, H., & Woods, B. K. S. (2018). An investigation into 3D printing of fibre reinforced thermoplastic composites. *Additive Manufacturing*, 22, 176–186.

Boros, R., Rajamani, P. K., & Kovacs, J. G. (2019). Combination of 3D printing and injection molding: overmolding and overprinting. *Express Polymer Letters*, 13(10), 889–897.

Camacho, D. D., Clayton, P., O'Brien, W. J., Seepersad, C., Juenger, M., Ferron, R., & Salamone, S. (2018). Applications of additive manufacturing in the construction industry-A forward-looking review. *Automation in Construction*, 89, 110–119.

Campbell, I., Bourell, D., & Gibson, I. (2012). Additive manufacturing: rapid prototyping comes of age. *Rapid Prototyping Journal*, 18(4), 255–258.

Chávez, F. A., Quiñonez, P. A., & Roberson, D. A. (2021). Hybrid metal/thermoplastic composites for FDM-type additive manufacturing. *Journal of Thermoplastic Composite Materials*, 34(9), 1193–1212. https://doi.org/10.1177/0892705719864150.

Daminabo, S. C., Goel, S., Grammatikos, S. A., Nezhad, H. Y., & Thakur, V. K. (2020). Fused deposition modeling-based additive manufacturing (3D printing): techniques for polymer material systems. *Materials Today Chemistry*, 16, 100248.

Deshwal, S., Kumar, A., & Chhabra, D. (2020). Exercising hybrid statistical tools GA-RSM, GA-ANN and GA-ANFIS to optimize FDM process parameters for tensile strength improvement. *CIRP Journal of Manufacturing Science and Technology*, 31, 189–199. https://doi.org/10.1016/j.cirpj.2020.05.009.

Dilberoglu, U. M., Gharehpapagh, B., Yaman, U., & Dolen, M. (2017). The role of additive manufacturing in the era of industry 4.0. *Procedia Manufacturing*, 11, 545–554.

Dilberoglu, U. M., Gharehpapagh, B., Yaman, U., & Dolen, M. (2021). Current trends and research opportunities in hybrid additive manufacturing. *International Journal of Advanced Manufacturing Technology*, 113(3–4), 623–648. https://doi.org/10.1007/s00170-021-06688-1

Dornelas, P. H. G., Santos, T. G., & Oliveira, J. P. (2022). Micro-metal additive manufacturing-state-of-art and perspectives. *The International Journal of Advanced Manufacturing Technology*, 122(9–10), 3547–3564.

Elhabashy, A. E., Wells, L. J., Camelio, J. A., & Woodall, W. H. (2019). A cyber-physical attack taxonomy for production systems: a quality control perspective. *Journal of Intelligent Manufacturing*, 30, 2489–2504.

Flynn, J. M., Shokrani, A., Newman, S. T., & Dhokia, V. (2016). Hybrid additive and subtractive machine tools-Research and industrial developments. *International Journal of Machine Tools and Manufacture*, 101, 79–101.

Hasanov, S., Alkunte, S., Rajeshirke, M., Gupta, A., Huseynov, O., Fidan, I., Alifui-Segbaya, F., & Rennie, A. (2022). Review on additive manufacturing of multi-material parts: progress and challenges. *Journal of Manufacturing and Materials Processing*, 6(1), 4. https://doi.org/10.3390/jmmp6010004

Heeling, T., Cloots, M., & Wegener, K. (2017). Melt pool simulation for the evaluation of process parameters in selective laser melting. *Additive Manufacturing*, 14, 116–125.

Jihong, Z. H. U., Han, Z., Chuang, W., Lu, Z., Shangqin, Y., & Zhang, W. (2021). A review of topology optimization for additive manufacturing: status and challenges. *Chinese Journal of Aeronautics*, 34(1), 91–110.

Jiménez, A., Bidare, P., Hassanin, H., Tarlochan, F., Dimov, S., & Essa, K. (2021). Powder-based laser hybrid additive manufacturing of metals: a review. *International Journal of Advanced Manufacturing Technology*, 114(1–2), 63–96. https://doi.org/10.1007/s00170-021-06855-4.

Kumar, P., Gupta, P., & Singh, I. (2022a). Parametric optimization of FDM using the ANN-based whale optimization algorithm. *AI EDAM*, 36, e27. https://doi.org/ DOI: 10.1017/S0890060422000142.

Kumar, P., Gupta, P., & Singh, I. (2022b). Performance analysis of acrylonitrile-butadiene-styrene-polycarbonate polymer blend filament for fused deposition modeling printing using hybrid artificial intelligence algorithms. *Journal of Materials Engineering and Performance*, 32, 1924–1937. https://doi.org/10.1007/s11665-022-07243-z.

Li, C. Z., Zhao, Y., Xiao, B., Yu, B., Tam, V. W. Y., Chen, Z., & Ya, Y. (2020). Research trend of the application of information technologies in construction and demolition waste management. *Journal of Cleaner Production*, 263, 121458.

Li, L., Haghighi, A., & Yang, Y. (2018). A novel 6-axis hybrid additive-subtractive manufacturing process: design and case studies. *Journal of Manufacturing Processes*, 33, 150–160.

Liravi, F., & Toyserkani, E. (2018). A hybrid additive manufacturing method for the fabrication of silicone bio-structures: 3D printing optimization and surface characterization. *Materials & Design*, 138, 46–61.

Liu, R., Wang, Z., Sparks, T., Liou, F., & Newkirk, J. (2017). Aerospace applications of laser additive manufacturing. In *Laser Additive Manufacturing* (pp. 351–371). Elsevier.

Matouš, K., Geers, M. G. D., Kouznetsova, V. G., & Gillman, A. (2017). A review of predictive nonlinear theories for multiscale modeling of heterogeneous materials. *Journal of Computational Physics*, 330, 192–220.

McKenzie, J. (2018). *Hybrid Additive Manufacturing of Multiphase Materials*. North Carolina Agricultural and Technical State University. University ProQuest Dissertations Publishing.

Meda, M. (2021). *Direct Writing of Printed Electronics through Molten Metal Jetting*. Rochester Institute of Technology. ProQuest Dissertations Publishing.

Mehrpouya, M., Dehghanghadikolaei, A., Fotovvati, B., Vosooghnia, A., Emamian, S. S., & Gisario, A. (2019). The potential of additive manufacturing in the smart factory industrial 4.0: a review. *Applied Sciences*, 9(18), 3865.

Mellor, S., Hao, L., & Zhang, D. (2014). Additive manufacturing: a framework for implementation. *International Journal of Production Economics*, 149, 194–201.

Merklein, M., Junker, D., Schaub, A., & Neubauer, F. (2016). Hybrid additive manufacturing technologies-an analysis regarding potentials and applications. *Physics Procedia*, 83, 549–559.

Ngo, T. D., Kashani, A., Imbalzano, G., Nguyen, K. T. Q., & Hui, D. (2018). Additive manufacturing (3D printing): a review of materials, methods, applications and challenges. *Composites Part B: Engineering*, 143, 172–196.

Parupelli, S. K., & Desai, S. (2020). Hybrid additive manufacturing (3D printing) and characterization of functionally gradient materials via in situ laser curing. *The International Journal of Advanced Manufacturing Technology*, 110, 543–556.

Pawlowski, A. E., Splitter, D. A., Muth, T. R., Shyam, A., Carver, J. K., Dinwiddie, R. B., Elliott, A. M., Cordero, Z. C., & French, M. R. (2017). Producing hybrid metal composites by combining additive manufacturing and casting. *Advanced Materials and Processes*, 175(7), 16–21.

Pecullan, S., Gibiansky, L. V, & Torquato, S. (1999). Scale effects on the elastic behavior of periodic andhierarchical two-dimensional composites. *Journal of the Mechanics and Physics of Solids*, 47(7), 1509–1542.

Petrovic, V., Vicente Haro Gonzalez, J., Jordá Ferrando, O., Delgado Gordillo, J., Ramón Blasco Puchades, J., & Portolés Griñan, L. (2011). Additive layered manufacturing: sectors of industrial application shown through case studies. *International Journal of Production Research*, 49(4), 1061–1079.

Popov, V. V., & Fleisher, A. (2020). Hybrid additive manufacturing of steels and alloys. *Manufacturing Review*, 7, 9. https://doi.org/10.1051/mfreview/2020005.

Pragana, J. P. M., Sampaio, R. F. V., Bragança, I. M. F., Silva, C. M. A., & Martins, P. A. F. (2021). Hybrid metal additive manufacturing: a state-of-the-art review. *Advances in Industrial and Manufacturing Engineering*, 2(January), 100032. https://doi.org/10.1016/j.aime.2021.100032.

Rajan, K., Samykano, M., Kadirgama, K., Harun, W. S. W., & Rahman, M. M. (2022). Fused deposition modeling: process, materials, parameters, properties, and applications. *The International Journal of Advanced Manufacturing Technology*, 120(3–4), 1531–1570.

Reith, M., Franke, M., Schloffer, M., & Körner, C. (2020). Processing 4th generation titanium aluminides via electron beam based additive manufacturing-characterization of microstructure and mechanical properties. *Materialia*, 14, 100902.

Salifu, S., Desai, D., Ogunbiyi, O., & Mwale, K. (2022). Recent development in the additive manufacturing of polymer-based composites for automotive structures-A review. *The International Journal of Advanced Manufacturing Technology*, 119(11–12), 6877–6891.

Sefene, E. M., Hailu, Y. M., & Tsegaw, A. A. (2022). Metal hybrid additive manufacturing: state-of-the-art. *Progress in Additive Manufacturing*, 7(4), 737–749. https://doi.org/10.1007/s40964-022-00262-1.

Sepasgozar, S. M. E., Shi, A., Yang, L., Shirowzhan, S., & Edwards, D. J. (2020). Additive manufacturing applications for industry 4.0: a systematic critical review. *Buildings*, 10(12), 231.

Sidambe, A. T. (2014). Biocompatibility of advanced manufactured titanium implants-A review. *Materials*, 7(12), 8168–8188.

Silva, M. R., Domingues, J., Costa, J., Mateus, A., & Malça, C. (2019). Study of metal/polymer interface of parts produced by a hybrid additive manufacturing approach. *Applied Mechanics and Materials*, 890, 34–42. https://doi.org/10.4028/www.scientific.net/amm.890.34.

Singamneni, S., Yifan, L. V, Hewitt, A., Chalk, R., Thomas, W., & Jordison, D. (2019). Additive manufacturing for the aircraft industry: a review. *Journal of Aeronautics & Aerospace Engineering*, 8(1), 351–371.

Sun, C., Wang, Y., McMurtrey, M. D., Jerred, N. D., Liou, F., & Li, J. (2021). Additive manufacturing for energy: a review. *Applied Energy*, 282, 116041.

Uçak, N., Çiçek, A., & Aslantas, K. (2022). Machinability of 3D printed metallic materials fabricated by selective laser melting and electron beam melting: a review. *Journal of Manufacturing Processes*, 80, 414–457.

Yamazaki, T. (2016). Development of a hybrid multi-tasking machine tool: integration of additive manufacturing technology with CNC machining. *Procedia Cirp*, 42, 81–86.

Zhu, Z., Dhokia, V. G., Nassehi, A., & Newman, S. T. (2013). A review of hybrid manufacturing processes - State of the art and future perspectives. *International Journal of Computer Integrated Manufacturing*, 26(7), 596–615. https://doi.org/10.1080/0951192X.2012.749530.

Chapter 7

Hybrid additive manufacturing of composite materials

Shenbaga Velu P
Vellore Institute of Technology

Rajesh Jesudoss Hynes N
Opole University of Technology
Mepco Schlenk Engineering College

Tharmaraj R
SRM Institute of Science and Technology

7.1 INTRODUCTION OF POLYMER MATRIX COMPOSITES

Polymer matrix composites are multiphase materials comprising a polymer intercellular model and an uninterrupted or discontinuous filler ingredient [1]. The qualities of the polymer pattern and a combined mixture of apparatus form the effects made on the intercellular substance [2], sandwiched material, the intersection, and scattered wadding substance in the polymer model [3], and the geometric organisation and preparation of the sandwiched substantial in between polymer grid [4]. Composite materials with unusual mechanical properties and anisotropic power-driven electrical and thermic conductivity can be designed and manufactured by changing matrix and filler material properties [5–9]. These substances are of great need to the technological and systematic groups of organisations due to their valuable perspective. Because of this, many distinct classes of manufacturing processes have been put into use.

Traditional manufacturing techniques of polymer intercellular models with uninterrupted wadding material generate high standards and a huge range of samples, but often include many costly production steps, such as mould creation, polymer adhesive, impregnation, preservation, and/or after-treatment methods [10]. In contrast, standard production methods for the polymer intercellular model with intermittent wadding material include mould ejection [11], and injection [12] involves injecting a combination of the liquid polymer model and sandwiched material between hollow holes. These technologies can produce more complicated specimen geometries and are cheaper than conventional methods for mixed

DOI: 10.1201/9781003406488-7

materials with uninterrupted filler elements because of the combined composition of a polymer and a discontinuous filling material that proceed into packed crevices and sections, computerised corners, and automated mould procedures that swiftly expel completed samples before recycling the mould. However, they provide little domination over the dimensional organisation and positioning of the sandwiched material and the alignment of the filler material, and each specimen geometry requires a unique mould.

Additive fabrication procedures include fused filament fabrication (FFF), direct ink writing (DIW), fused deposition modelling (FDM), and stereolithographic, which may create complex specimen shapes without a mould [13]. There is minimal leftover scrap in traditional manufacturing procedures [14]. We can use an additive fabrication model to make polymer intercellular composite models with exclusive qualities that permit previously unattainable mediums through geometric arrangements or positioning/orienting interrupted padding mediums.

7.2 PMCM MANUFACTURE METHODS

7.2.1 FMA method

Filler substances differ in configuration and dimensional proposition. Most materials used on a large scale are fibrous and may be made of carbon, Kevlar, or glass fibre tow in either continuous or chopped form [15]. Microscale fillers [16] include finely chopped microfibres, microrods, spherical and nonspherical particle powders, and microrods. Nanoscale filler materials include single- and multi-walled carbon nanotubes, nanofibres, graphene nanoplatelets, and spherical particle powders [17].

There are three primary methods by which filler material is incorporated into a polymer matrix:

1. integrating persistent macrofibre strands into the polymer intercellular model [18]
2. Processing a liquid polymer or uninterrupted nanofibrous material into a continuous fibre prior to embedding it in the polymer pattern, which has been consummated by the use of melt spinning [19], electrospinning [20], and wet spinning [21, 22], and directly spreading intermittent filler elements into the polymer pattern [23]. The processing methods orient continuous fibres using geometrical and mechanical (tension) forces, such as rollers in an FFF or FDM printer's extrusion nozzle. In contrast, the third produces unsystematically modified and scatteredly sandwiched material in the polymer pattern.

The matrix's filler material's dimensional organisation and positioning determine the polymer matrix composite specimen's properties. It is common knowledge that placing micron- and nanometer-sized sandwiched material aligned with the computerised loads strengthens the material and increases its electrical and thermal conductivity by lowering the percolation threshold [24].

7.2.2 Mould casting

In Figure 7.1, a liquid polymer adhesive and discontinuous wadding material are injected into a 3D mould chamber made of glass, metal, or rigid polymer. The polymer resin determines the shape of the material specimen and the sandwiched material combined, conforming to the chamber of the three-dimensional mould (Figure 7.1a). The external field transducers' force controls the material specimen's microstructure, which coordinates the padding material spread in the polymer exudate into a described inclination and position (Figure 7.1b). The medium is fixed in the polymer intercellular space by the liquid polymer resin hardening or curing before attaining rigidity from the mould chamber (Figure 7.1c). We employ mould ejection

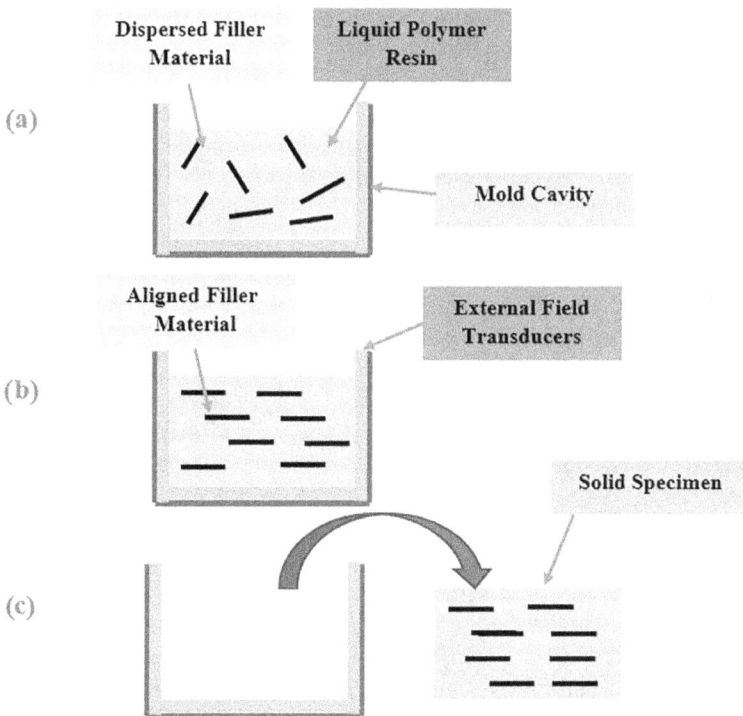

Figure 7.1 (a) Diagrammatic model of a representative mould assembly framework in which (a) a combination of aqueous polymer adhesive and scattered sandwich material is inserted into a three-dimensional mould chamber, (b) an outermost field that adjusts the padding material, and (c) the polymer adhesive that solves the rigid trail is separated from the mould chamber.

as a basis of comparison for additive manufacturing procedures. Extrusion and stereolithography (SLA) are additive manufacturing methods.

Khan et al. [25] aligned 0.5 wt% nanoparticles in a polymeric matrix within a fibreglass mould using DC electric fields. UV light and ozone treatment made MWCNT surfaces hydrophilic, improving dispersion and matrix adherence. They discovered that the MWCNT weight percentage had an effect on the electrical conductivity and that this effect was parallel to the direction of alignment. Ladani et al. [26] employed mould casting to align up to 1.6 wt% using an AC electric current. Two lightweight carbon pre-impregnated strips that also serve as probes are used to create a 2-mm-thick rectangle chamber that contains CNF in a polymeric matrix. The composite structure samples were constrained with mould form and CNF alignment time; however, they showed that conductance increased with fillers wt% and alignment by decreasing the relative permeability. Alignment is especially important for small CNF wt% compared to large CNF wt% since it becomes more difficult to construct a penetrable network of carbon fibres with decreasing wt%.

Erb et al. [27] created layered hydrogel composites with aligned aluminium oxide platelets using mould casting and an external magnetic field. To make the aluminium oxide platelets ferromagnetic, they covered them with metallic nanoparticles. This was followed by spinning a peculiar magnetic field over the nonstick coating formation. Besides mounting mould cavity layers, the hydrogel matrix self-shapes in programmable patterns.

Greenhall et al. [28] have used the ultrasound sector in the gauge length of a dogbone-shaped adhesive mould to make large pieces of urethane matrix with far more than 10 wt% oriented MWCNTs. To get to such a high weight-to-volume ratio, scientists focused a mass of matrix material in which the MWCNTs no longer resided using sonic focusing and then eliminated the extra matrix material. This method avoids the high density and dispersing problems caused by the large weight ratio of filler material in the matrix material. They demonstrated that, in comparison to the virgin matrix, the polymer matrix composite material's ultimate tensile strength improved by 73% due to the ultrahigh weight percentage of MWCNTs.

7.2.3 Extrusion fabrication methods

Extrusion creates a three-dimensional structure with discrete layers by selectively depositing an aqueous polymer with a ceaseless or scattered stopgap ingredient through an outlet. Figure 7.2 shows a polymer matrix and continuous filler FFF system. A heated extruder nozzle feeds uninterrupted wadding material (black) and polymer intercellular (yellow) filament wires. As it leaves the outlet and sublimates on the frame top, the polymer intercellular material melts and envelops the uninterrupted wadding material. Adjusting the disposition of the extruder outlet or construction plate to layer-by-layer sediment of sandwich material and polymer intercellular material yields a three-dimensional specimen.

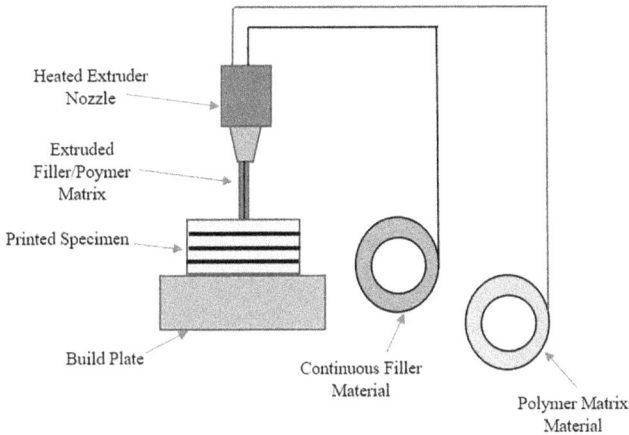

Figure 7.2 The filament fills into a warmed extruder outlet and particularly sediments on a construction plate to preprint a 3D test trail layer by layer.

A computer-controlled extruder nozzle heats, melts, extrudes, and intentionally deposits a polymer fibre on a construction layer into a customised 3D symmetry in FFF. Since it requires the filament to melt and resolidify, thermoplastics such as polypropylene, nylon, ABS, PLA, and polyamide are your only options for filament materials (PP). FFF-printed polymer matrix composite products often use continuous Kevalar, glass, carbon as filler, short-chopped carbon fibres, and CNFs.

Thermoplastic filaments can have extruded uninterrupted filler material and be deposited into them using a current FFF printer by inserting them with the filler material beforehand. Shear intensity in the intersecting extruder outlet shape modifies discontinuous padding material as it settles and hardens on a construction plate in FFF systems. The PLA composite materials utilised by Ferreira et al. [29] (1.75 mm filament diameter, 60 μm microfibre length) were printed using FFF, and they featured aligned carbon microfibres. or tensile and shear testing, they produced $165 \times 19 \times 3.3$ mm dogbone and $200 \times 25 \times 4.8$ mm rectangular composite material specimens $0°$, $90°$, and $\pm45°$ layer orientations, 50 mms^{-1} print rate, and a 100% rectilinear backfill. They found that adding 15 wt% carbon microfibres to virgin PLA composites increased exemplar rigidity by 220% in the printing indication.

Using FFF, the polymer intercellular and the wadding can be combined in the extruder nozzle before being deposited concurrently on the build plate. Hou et al. [30] made 60 mm× 60 mm× 15 mm composite sandwich structures using an FFF arrangement with individual fibre spools of uninterrupted Kevlar fibres (solidity$=1{,}440$ kgm^3) and PLA adhesive (1.75 mm diameter) in a particular extruder outlet. The mechanical properties of specimens produced by printing process parameters, including layer

(0.1–0.5 mm), cell dimension (9–13 mm), and sandwiched material density proportion (2%–11.5%), were examined. With increasing filler material volume per cent, the composite material reached 17.17 MPa at 11.5 vol%. Heidari-Rarani et al. [31] built an extruder nozzle that could attach to commercial FFF printers and extrude PLA (1.75 mm dia) and continuous carbon fibre (7 mm diameter). They created rectangular PLA composite material specimens with a volume fraction of 28.2% carbon fibre, a print speed of 20 mms^{-1}, and a layer thickness of 300 μm. They recommended a minimum interval of 0.4–0.5 mm between consecutive preprint orders to prevent carbon fibre cracking in test trail radii.

7.2.4 SLA fabrication method

SLA technologies, which are used in additive manufacturing, work by selectively projecting UV illumination into a barrel containing fluorescent polymer adhesive to polymerise a two-dimensional shape. Figure 7.3 depicts a

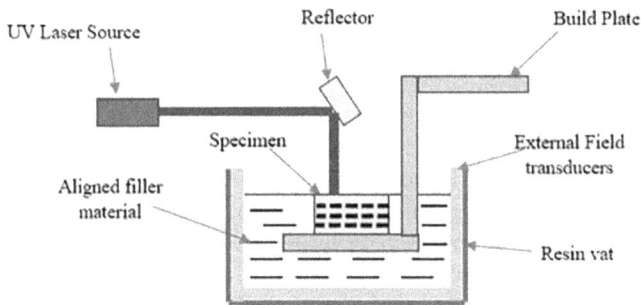

(a) Laser stereolithography

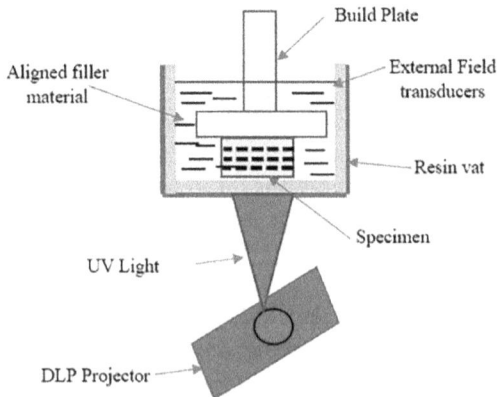

(b) Projection stereolithography

Figure 7.3 Diagrammatic of a complex (a) laser and (b) DLP prediction SLA additive fabrication arrangement.

complex digital light processing (DLP) forecast SLA configuration. In all situations, fluorescent adhesive is combined using an interrupted wadding and forged into an adhesive barrel. Exterior arena transducers shown in red arrange the wadding substantially preceding an ultraviolet illumination source, particularly refining a 2D surface of photopolymer adhesive. The employed fine film of photopolymer adhesive organises itself into a computerised construction platter, allowing the construction of a three-dimensional trail layer by layer.

Llewellyn Jones et al. [32] used a cast-off laser-based SLA and an ultrasonic tendency field to make single-layer samples of photopolymer composite materials with synchronised glass microfibres. A motion-controlled laser beam selectively cured the photopolymer after parallel-oriented acoustic transducers arranged the glass fibres scattered throughout the resin into lines with equal spacing. They struggled to control layer thickness and produce multilayer samples due to the absence of a build plate in their arrangement.

UV light forecast SLA technologies remedy a given material layer at a time, whereas laser-based SLA systems map photosensitive areas to cure, slowing down fabrication. Martin et al. [33] used a rotating magnetic field and projection-based SLA to position 15% iron oxide-coated ceramic microplatelets into 90-micron structures. Using electromagnetic solenoids underneath and around the photopolymer reservoir, they successively and sequentially generated photosensitive polymer layers and changed magnetisation orientation to achieve the filler material architecture.

Greenhall and Raeymaekers [28] used a perception SLA and an ultrasonic light field in octal acrylic storage with four adjacent ultrasound transducers. They claimed that the material developed anisotropic electrical conductivity because of the penetrable matrix formed by the rows of oriented fibres. Following this, Niendorf and Raeymaekers [34] utilised the fabrication method developed by Greenhall and Raeymaekers to 3D print carbon microfibre composite materials measuring 10.00 mm× 5.00mm×0.75 mm in a single layer. They found that fabrication factors such as ultrasonic transducer input voltage, filler material wt%, and ultrasound transducer distance can disconnect macro- and microscale orientation.

7.2.5 Direct ink writing

In DIW, sector sedimentation, an aqueous polymer intercellular compound ink, sediments, extrudes, and deploys to create a 3D material illustration. In contrast to FFF, which melts thermoplastic filaments, DIW extrudes colloids, nanoparticles, thermosets, and organic materials from liquid ink reservoirs. To pass through the extruder outlet, DIW inks must be compressive but viscous enough to maintain shape after deposition.

i. DIW Combined with a Shear Force Field

Compton and Lewis used epoxy ink and several shear-aligned fillers to 3D preprint integral blended materials. For structural support, they added silicon carbide whiskers, nanoclay platelets, and carbon microfibres. Then, using specimens with cell wall thicknesses ranging from 200 to 350 μm and diameters varying from 88 to 20 mm, they employed a DIW printer to produce traditional symmetric designs modulated by balsa wood. Gladman et al. [35] created natural-material-inspired polymer matrix composite structures of 57.5×30 mm using DIW. They made 600-μm hydrogel filaments containing shear-arranged fibrous material that resemble plant tissue walls. The asymmetrical rigidity and swelling in reaction to exterior stimuli enabled the design and 3D preprint of structure-modifying materials with complicated geometry. The shear constrain fields with DIW-aligned carbon microfibres in an elastomeric material by Lewicki et al. [36]. Si nanoparticles and carbon microfibres improve resin shear disintegration, which is necessary for consistent extrusion and nozzle clogging.

ii. DIW Combined with a Magnetic Field

In 15 mm× 15mm×15 mm polyurethane specimens, Kokkinis et al. [37] aligned 1% of weight alumina particles coated with metal nanoparticles to make them saturation magnetisation. Ink was deposited on a build platform via several extruder nozzles, allowing for the use of many ink formulas in a single print. Then the work level proceeds through a pivoting attractive field (40 mT, 500 rpm) to arrange the particles in a positive way.

iii. DIW Combined with an Ultrasonic Wave Field

Friedrich et al. [38] printed 350–750 μm broad compound material trails, including an epoxy cellular material with organised glassy microspheres, using DIW in conjunction with an ultrasonic wave. Compared to shear force field alignment approaches, inserting ultrasonic transducers complicates extruder nozzle design, but the minimal pressure zones of the ranking ultrasound wave employ filler material alignment and arrangement. Friedrich et al. [38] investigated the effects of preprinting factors sandwiched substantial size proportion, ultrasound transducer input voltage variation [0–50 Vpp], preprint rate [1–20 mms^{-1}], ink characteristics, acetone injection, and ultrasonic wave range boundaries on filler material geometrical arrangement and ink shape retention.

To produce epoxy fibre (up to 25 mm in length) with various sandwiched materials, particularly organised fillers, including other rows by their material qualities Collino et al. [39] employed DIW with an ultrasonic wave field ranges (2.04–3.32 MHz operating frequency). Wadsworth et al. [40] created 80mm×10mm×1 mm compound material trails using a photoresist and rows comprising arranged carbon microfibres coated with nickel utilising DIW with a 2.15–6.35 mms^{-1} design rate and an ultrasonic trend field of various working frequencies (1.0–2.0 MHz). When depositing the filament onto the build plate, the carbon microfibres are aligned using acoustic transducers attached to a modified extruder nozzle. Optical imaging and electrical conductivity studies proved that the microfibres were aligned on macroscopic and microscopic scales after production.

7.2.6 Compatibility between fabrication and filler material alignment methods

Mould cavity aligns filler material with electrical, magnetisation, or acoustic waves. Maximum stress orientation material arrangement is tough to use in mould casting owing to poor substantial movement direction control. FFF extrusion aligns filler material using perfunctory force or tear force fields. FFF employing electrical, magnetisation, or acoustic wave padding material alignment procedures has not been explored. Because the nozzle traces the specimen's 3D contour in FFF procedures, the filler material has little time to align due to liquid thermoplastic polymer resins' high viscosity. All techniques for aligning filler materials using an external field must take into account the solubility of the resin polymer. This is so that the peripheral field's driving force can outweigh the drag force caused by a viscous fluid, which not only prevents filler material alignment but also keeps it from settling or precipitating [41].

In the orientation of filler material, in addition to a maximum stress field, DIW manufacturing techniques use resin matrix substances that are less viscous and print at a lesser rate than FFF techniques. Ideally, a specimen can be printed first, and the filler material could then be aligned using an electric field. Alternatively, electrodes may be incorporated into the DIW nozzle to orient the padding material while printing. The enormous vat of stationary photopolymer resin used in SLA production techniques and its few moving parts make it possible to integrate several transducers to produce peripheral electrical, magnetisation, or acoustic waves. SLA relies on matrix flow to align the filler material and does not integrate with a maximum stress field since the photopolymer is immobile. Moreover, no studies have been done on SLA with mechanical alignment.

7.2.7 Specimen size

The size of the specimen may be limited by the manufacturing or padding material alignment process. The flow of specimen orientation padding material during manufacture using mechanical and shear force mechanisms. Hence, the orientation method of the padding material is not constrained by the production method, only the sample shape. Specimen geometry is typically limited by the placement instead of the fabrication techniques when electrical, electromagnetic, or acoustic radiation filler material orientation is being used. This is because the shape of the specimen raises capacitances, transducer capacity, and running capacity.

7.2.8 Resolution

The specimen geometry resolution depends on the precision and viscosity of the mould since mould casting needs molten polymer resin to take the form of the mould while solidifying. This implies that precise structures in the mould won't be formed properly, whereas epoxy has a stiffness that prevents it from flowing into complex cavities.

7.2.9 Suitable filler materials

Composite polymer matrix materials' filler components vary in terms of wt%, aspect ratio, and substance, but their orientation and production processes are compatible. Extruder nozzles must instantly align continuous filler material for mechanical alignment methods. Shear force fields exert pressure transverse to the shearing plane on padding materials that have a large aspect ratio [42]. Viscosity and drag force increase as filler material weight percentage increases, necessitating an increase in external field strength. Hence, when using a magnetic, electric, or ultrasonic upsurge field, the padding material wt% is often low.

7.3 METAL/POLYMER JOINTS

Because of their minimum weight, wear, corrosion resistance, thermic and electrical adherence, and model flexibility, plastic-based materials are widely used in the automobile, aircraft, and electronic sectors [43]. When joining metal and plastic together, there are frequently challenges to overcome, and the blended nature is not entirely learned. Metal and plastic materials have different mechanic and structural qualities, and hybrid medium has limited joining methods [44].

Friction stir welding (FSW) is one of the technologies that has made it possible to weld incompatible metals and polymers. The non-consumable hold is kept in the medium at the corner of the mould to be produced in FSW. The interior top of the instrument is permitted to be fixed to the surface of the sample. The rotational speed of the tool is specified, and it is horizontal across the lap joint. Localised softening and elastic mould deformation in the specimen material are seen as a result of the tool's interaction with the workpiece. The tool presses the material and the pin together from top to bottom and front to back, using rotational and linear motion to create the joint. Aluminium alloys, steel, copper, titanium, complex metal matrices, magnesium, and thermoplastics are just some of the numerous materials that have been successfully welded with this method because of its versatility.

The FLW process looks like the FSW process. The vital variaton lies between the FLW and the FSW is that the FLW equpiment doesn't have a stiff pin. Because of this, the initial procedure of the rotation equipment is not to make materials go along with the stirring pin but to force and apply heat to the metal specimen. The plastic component receives heat through conductive flow from the warmed metal compound, softening the elastic mould materials in a constrained area near the interface. After the molten plastic mould has solidified under the huge amount of pressure applied by the pressed metal complex, a bond between the metal and plastic mould can be formed.

The interface between aluminium and polyphenylene sulphide includes voids and pores, as shown by the work of Ratanathavorn et al. [45]. Along the boundary line, it was demonstrated that one could observe the formation of pores and cavities in the material. Also visible is a combination of metal and thermoplastic particles.

Shahmiri et al. [46] studied combining elements made of Al/PP and found that the weld nugget was of high quality when the rotational speed was 800 rpm rather than 1,200 rpm. The larger the load-bearing region of an attachment, the more prominent the weld nugget. The base aluminium and heat-affected zone exhibit prolonged cold-rolled grains. In the thermo-mechanically affected zone, carried grains are distorted and stretched in the path of the pins' stiffening action. Even though temperature and strain are insufficient to cause recrystallisation, in the stir zone, aluminium was treated to severe strain and high temperature, resulting in rolled grains recrystallising into excellent equiaxed grains. Huang et al. [47] employed a sealed thread pinhead with three faces to join PEEK and Al linen while preserving the metal in the top cavity. At the minimum translational rate, when the heat input is substantially greater, an excess of material can be observed, including the pores and voids. As the varying translational velocity tends to extend, the observable surge decreases. After a particular level, however, the metal fragments and minute particles are observed to be non-uniformly mixed in the stir zone. Liu et al. [48] welded nylon-6 and Al alloy using FLW. It was noticed that the existence of bubbles along the interconnection of metal and polymer explains a proportion of bubble dimensions to overlapped region dimensions at the contact. Varying welding speeds and rotational speeds cause variations in nylon thickness. Buffa et al. [49] utilised friction overlap welding behind the extruded polymer into an aluminium sheet. The polymer was heated by applying frictional heat generated by the FLW tool. As the machine is moved forward, the composite behind it extrudes into the metal sheet through the holes. These composite extrusions mechanically interlock with the metal upon cooling. Hole spacing was calculated as a multiple of the diameter of the drilled hole.

i. Tensile strength

According to Ratanathavorn et al. [45], an area filled with chips that are in undeviating proximity to the metallic trench has more stablility than the surrounding elastic mould and pores. Since metallic fragments are added to the stir zone, the mechanical interlocking is better than when the stir zone was made of only pure thermoplastics. Smaller holes increase strength as translational speed increases.

Shahmiri et al. [46] reported that 800 rpm generated the least heat, coinciding with the highest wear firmness. Because of the greater nugget region and smaller gap, this occurred.

The highest shear strength and elongation were observed at 50 mm/min translational speed, according to Huang et al. [50]. There was less of an overflow; the resolidified polymer included an even distribution of metal and substantial anchors formed on AS and RS, all of which contributed to these results. The production of bubbles, which led to a decrease in load-resisting area, contributed to the sample's strength at a rate of 30 mm/min.

ii. Hardness

Shahmiri et al. [46] found that the formation of subgrains due to partial recrystallisation causes the hardness of HAZ to decrease. Recrystallisation causes the development of finely equiaxed grains, which increases the hardness of SZ. Due to the fragmentation and incorporation of the particles of these tiny equiaxed aluminium grains into the polymer matrix, the weld nugget region exhibits an improvement in hardness. Due to the thermal deterioration of the polymer, the polymer's hardness decreased along the interface. The temperature rises due to an increased rotating speed, leading to increased recrystallisation in the HAZ and grain growth in the SZ. It is possible to observe that the hardness rises as the rotational speed increases, which can be seen as the case.

7.4 SUMMARY

The ability to fabricate intricate PMC specimen geometries without the use of a mould makes additive manufacturing techniques such as FDM or FFF, SLA, and DIW significantly superior to traditional mould-casting fabrication techniques. It is possible to spatially organise and orient continuous or discontinuous filler material by employing mechanical force or an external electric, magnetic, shear force, or ultrasound wave field when additive manufacturing methods are combined with a filler material alignment method. This allows for the fabrication of polymer matrix composite materials with designer qualities that depend on the properties of and interaction between the matrix and filler material, as well as their spatial distribution and orientation in the matrix. These fabrication and filler material alignment procedures require precise tweaking, making bulk manufacture difficult. Compared to conventional fabrication procedures, their capacity to manufacture tailored polymer composite materials with designer material qualities shows promise.

REFERENCES

[1] Park, S. J., and Seo, M. K. (2011). Interface science and technology, In: *Interface Science and Composites*, 1st Edition, 18, Elsevier, Amsterdam, 501–629.

[2] Kalsoom, U., Nesterenko, P. N., and Paull, B. (2016). Recent developments in 3D printable composite materials, *RSC Adv.* 6: 60355.

[3] Ma, P. C., Siddiqui, N. A., Marom, G., and Kim, J. K. (2010). Dispersion and functionalization of carbon nanotubes for polymer-based nanocomposites: a review, Compos. *A Appl. Sci. Manuf.* 41: 1345–1367.

[4] Xie, X. L., Mai, Y. W., and Zhou, X. P. (2005). Dispersion and alignment of carbon nanotubes in polymer matrix: a review, Mater. *Sci. Eng. R Rep.* 49: 89–112.

[5] Joseph, K., Malhotra, S. K., Goda, K., and Sreekala, M. S. (2012). Advances in polymer composites: macro- and microcomposites - state of the art, new challenges, and opportunities, *Polym. Compos.* 1: 1–16.

[6] Lau, A. K. T., and Hui, D. (2002). Carbon nanotube based composites - A review, *Compos. B Eng.* 33: 263.

[7] Wickramasinghe, S., Do, T., and Tran, P. (2020). FDM-based 3D printing of polymer and associated composite: a review on mechanical properties, defects and treatments, *Polym.* 12: 1529.

[8] Khan, S. U., Pothnis, J. R., and Kim, J. K. (2013). Effects of carbon nanotube alignment on electrical and mechanical properties of epoxy nanocomposites, *Compos. A Appl. Sci. Manuf.* 49: 26–34.

[9] Fischer, J. E., Zhou, W., Vavro, J., Llaguno, M. C., Guthy, C., Haggenmueller, R., Casavant, M. J., Walters, D. E., and Smalley, R. E. (2003). Single wall carbon nanotube fibers extruded from super-acid suspensions: preferred orientation, electrical, and thermal transport, *J.Appl. Phys.* 93: 1910–2157.

[10] Tong, L., Mouritz, A. P., and Bannister, M. (2002). *3D Fibre Reinforced Polymer Composites*, Elsevier Science & Technology, Oxford, 1–241.

[11] Delmonte, J. (1990). Metal/polymer structural composites, In: *Metal/ Polymer Composites.* Springer, Boston, MA. https://doi.org/ 10.1007/978-1-4684-1446-2_6.

[12] Kim, E. G., Park, J. K., and Jo, S. H. (2001). A study on fiber orientation during the injection molding of fiber-reinforced polymeric composites: (Comparison between image processing results and numerical simulation), *J. Mater. Process. Technol.* 111: 225–232.

[13] Gibson, I., Rosen, D., and Stucker, B. (2015). *Additive Manufacturing Technologies: 3D Printing, Rapid Prototyping, and Direct Digital Manufacturing*, Springer, Berlin.

[14] Gao, W., Zhang, Y., Ramanujan, D., Ramani, K., Chen, Y., Williams, C. B., Wang, C. C. L., Shin, Y.C., Zhang, S., and Zavattieri, P. D. (2015). The status, challenges, and future of additive manufacturing in engineering, *CAD Comput. Aided Des.* 69: 65–89.

[15] Goh, G. D., Yap, Y. L., Agarwala, S., and Yeong, W. Y. (2019). Recent progress in additive manufacturing of fiber reinforced polymer composite, *Adv. Mater. Technol.* 4: 1800271.

[16] Sugama, T., and Gawlik, K. (2003). Nanoscale boehmite filler for corrosion- and wear-resistant polyphenylenesulfide coatings, *Polym. Polym. Compos.* 11: 161.

[17] Yunus, D. E., Sohrabi, S., He, R., Shi, W., and Liu, Y. (2017). Acoustic patterning for 3D embedded electrically conductive wire in stereolithography, *J. Micromech. Microeng.* 27: 045016.

[18] Kabir, S. M. F., Mathur, K., and Seyam, A. F. M. (2021). Maximizing the performance of 3D printed fiber-reinforced composites, *J.Compos. Sci.* 5: 136.

[19] Huang, Z. M., Zhang, Y. Z., Kotaki, M., and Ramakrishna, S. (2003). A review on polymer nanofibers by electrospinning and their applications in nanocomposites, *Compos. Sci. Technol.* 63: 2223–2253.

[20] Pötschke, P., Brünig, H., Janke, A., Fischer, D., and Jehnichen, D. (2005). Orientation of multiwalled carbon nanotubes in composites with polycarbonate by melt spinning, *Polym.* 46: 10355–10363.

[21] Zhang, S., Koziol, K. K. K., Kinloch, I. A., and Windle, A. H. (2008). Macroscopic fibers of well-aligned carbon nanotubes by wet spinning, *Small.* 4: 1217–1222.

[22] Li, Y. L., Kinloch, I. A., and Windle, A. H. (2004). Direct spinning of carbon nanotube fibers from chemical vapor deposition synthesis, *Science*, 304: 276–278.

[23] Moniruzzaman, M., and Winey, K. I. (2006). Polymer nanocomposites containing carbon nanotubes, *Macromolecules.* 39: 5194–5205.

[24] Wang, Q., Dai, J., Li, W., Wei, Z., and Jiang, J. (2008). The effects of CNT alignment on electrical conductivity and mechanical properties of SWNT/ epoxy nanocomposites, *Compos. Sci. Technol.* 68: 1644–1648.

[25] Khan, S.U., Pothnis, J. R., and Kim, J. K. (2013). Effects of carbon nanotube alignment on electrical and mechanical properties of epoxy nanocomposites, *Compos. A Appl. Sci. Manuf.* 49: 26–34.

[26] Ladani, R. B., Wu, S., Kinloch, A. J., Ghorbani, K., Zhang, J., Mouritz, A. P., and Wang, C. H. (2016). Multifunctional properties of epoxy nanocomposites reinforced by aligned nanoscale carbon, *Mater. Des.* 94: 554–564.

[27] Ladani, R. B., Wu, S., Kinloch, A. J., Ghorbani, K., Mouritz, A. P., and Wang, C. H. (2017). Enhancing fatigue resistance and damage characterisation in adhesively-bonded composite joints by carbon nanofibers, *Compos. Sci. Technol.* 149: 116–126.

[28] Greenhall, J., and Raeymaekers, B. (2017). 3D printing macroscale engineered materials using ultrasound directed self-assembly and stereolithography, *Adv. Mater. Technol.* 2: 1700122–1700129.

[29] Ferreira, R. T. L., Amatte, I. C., Dutra, T. A., and Bürger, D. (2017). Experimental characterization and micrography of 3D printed PLA and PLA reinforced with short carbon fibers, *Compos. B.* 124: 88–100.

[30] Hou, Z., Tian, X., Zhang, J., and Li, D. (2018). 3D printed continuous fibre reinforced composite corrugated structure, *Compos. Struct.* 184: 1005–1010.

[31] Heidari-Rarani, M., Rafiee-Afarani, M., and Zahedi, A. M., (2019). Mechanical characterization of FDM 3D printing of continuous carbon fiber reinforced PLA composites, *Compos. Part B.* 175: 107147.

[32] Llewellyn-Jones, T. M., Drinkwater, B. W., and Trask, R. S. (2016). 3D printed components with ultrasonically arranged microscale structure, *Smart Mater. Struct.* 25: 02LT01.

[33] Martin, J. J., Fiore, B. E., and Erb, R. M. (2015). Designing bioinspired composite reinforcement architectures via 3D magnetic printing, *Nat. Commun.* 6: 8641.

[34] Niendorf, K., and Raeymaekers, B. (2020). Quantifying macro- and microscale alignment of carbon microfibers in polymer-matrix composite materials fabricated using ultrasound directed self-assembly and 3D-printing, *Compos. Part A.* 129: 105713.

[35] Gladman, S. A., Matsumoto, E. A., Nuzzo, R. G., Mahadevan, L., and Lewis, J. A. (2016). Biomimetic 4D printing, *Nat. Mater.* 15: 413–418.

[36] Lewicki, J. P., Rodriguez, J. N., Zhu, C., Worsley, M. A., Wu, A. S., Kanarska, Y., Horn, J. D., Duoss, E. B., Ortega, J. M., Elmer, W., Hensleigh, R., Fellini, R. A., and King, M. J. (2017). 3D-printing of meso-structurally ordered carbon fiber/polymer composites with unprecedented orthotropic physical properties, *Sci. Rep.* 7: 43401.

[37] Kokkinis, D., Schaffner, M., and Studart, A. R. (2015). Multimaterial magnetically assisted 3D printing of composite materials, *Nat. Commun.* 6: 8643.

[38] Friedrich, L., Collino, R., Ray, T., and Begley, M. (2017). Acoustic control of microstructures during direct ink writing of two-phase materials, *Sens. Actuators Phys.* 268: 213–221.

[39] Collino, R. R., Ray, T. R., Fleming, R. C., Cornell, J. D., Compton, B. G., and Begley, M. R. (2016). Scaling relationships for acoustic control of two-phase microstructures during direct-write printing, *Extrem. Mech. Lett.* 8: 96.

[40] Wadsworth, P., Nelson, I., Porter, D. L., Raeymaekers, B., and Naleway, S. E. (2020). Manufacturing bioinspired flexible materials using ultrasound directed self-assembly and 3D printing, *Mater. Des.* 185: 108243.

[41] Roy, M., Tran, P., Dickens, T., and Schrand, A. (2019). Composite reinforcement architectures: a review of field-assisted additive manufacturing for polymers. *J. Compos. Sci.* 4: 1.

[42] Gunes, D. Z., Scirocco, R., Mewis, J., Vermant, J., and Nonnewton, J. (2008). Flow-induced orientation of non-spherical particles: effect of aspect ratio and medium rheology. *J. Non newto. fluid Mech.* 155: 39–50.

[43] Katayama, S., and Kawahito, Y. (2008). Laser direct joining of metal and plastic. *Scripta Mater,* 59: 1247–1250.

[44] Amancio-Filho, S. T., and Dos Santos, J. F. (2009). Joining of polymers and polymers-metal hybrid structures: recent developments and trends. *Polym. Eng. Sci.* 49: 1416–1476.

[45] Ratanathavorn, W., and Melander, A. (2015). Dissimilar joining between aluminium alloy (AA 6111) and thermoplastics using friction stir welding, *Sci. Technol. Weld. Join.* 20: 222–228.

[46] Shahmiri, H., Movahedi, M., and Kokabi, A. H. (2017). friction stir lap joining of aluminium alloy to polypropylene sheets, *Sci. Technol. Weld. Join.* 22: 120–126.

[47] Huang, Y., Meng, X., Xie, Y., Li, J., Si, X., and Fan, Q. (2019). Improving mechanical properties of composite/metal friction stir lap welding joints via a taper-screwed pin with triple facets, *J. Mater. Process. Technol.* 268: 80–86.

[48] Liu, F. C., Liao, J., and Nakata, K. (2014). Joining of metal to plastic using friction lap welding, *Mater. Des.* 54: 236–244.

[49] Buffa, G., Baffari, D., Campanella, D., and Fratini, L. (2016). An innovative friction stir welding-based technique to produce dissimilar light alloys to thermoplastic matrix composite joints, *Procedia Manuf.* 5: 319–331.

[50] Huang, Y., Meng, X., Wang, Y., Xie, Y., and Zhou, L. (2018). Joining of aluminum alloy and polymer via friction stir lap welding, *J. Mater. Process. Technol.* 257: 148–154.

Chapter 8

Printing a sustainable future

How additive manufacturing is revolutionising the fight against plastic waste

Jagdeep Kaur, Atri Rathore,
Tarveen Kaur, and Prabal Batra
Chandigarh University Gharuan

8.1 INTRODUCTION

The manufacturing environment has become more dynamic since the beginning of the twentieth century. One of the most significant driving forces behind this change is the study and creation of new manufacturing technologies, which enable small-scale production to be more inventive and cost-effective. Businesses are being compelled to reconsider and re-evaluate where and how they conduct their manufacturing activities due to the rise of additive manufacturing [commonly known as three-dimensional (3D) printing] as a direct production technology. When compared to logical manufacturing and developmental mass production techniques, additive manufacturing technologies define AM as the "process to merge materials by stratification to manufacture objects from three-dimensional (3D) model data utilising CAD software." AM offers significant sustainability advantages while enabling the development of value chains that are shorter, smaller, more localised, and more collaborative. Three benefits of this technology stand out from the numerous significant advantages it offers:

1. By adapting manufacturing techniques and goods for additive manufacturing (AM), advantages can be derived during both the production and consumption stages.
2. The life of products can be extended via technological methods like repair, re-manufacturing, and restoration, in addition to more sustainable socio-economic patterns like improved person-product affinities and closer connections between producers and consumers.
3. Value chains have been reorganised to incorporate more locally focused manufacturing, quicker and more straightforward supply chains, creative distribution strategies, and new collaborations [1].

DOI: 10.1201/9781003406488-8

Despite these potential benefits, AM's sustainability has not been adequately examined. While it might be a catalyst for greater industrial sustainability, its effects on the industrial system could lead to different outcomes where less eco-efficient specialised manufacturing, customer expectations for customised items, and a higher rate of product obsolescence work together to raise resource consumption. Yet, the impacts of its application to the industrial system might lead to a different scenario in which less environmentally friendly localised manufacturing, consumer demands for customised goods, and a higher rate of product obsolescence work together to contribute to rising resource consumption. The sustainability implications of AM have so far either been examined widely or indepth by researchers who have focused on the issue of material and energy usage.

The purpose of this study is to begin to sort through the issues that arise at the intersection of these fields as a new area of study, where the impact of AM on sustainability is unknown. The topic to be raised is as follows: How could additive manufacturing assist in more ecologically friendly production and consumption practises?

It is necessary to view the topic through the lens of industrial sustainability in order to fully grasp the significance of AM for improving the sustainability of industrial systems [2]. These systems are "complicated with a variety of participants on a global stage where various business entities engage with one another in complex, interconnected value chains, trading data, goods (raw materials, components, and products), services, and, of course, money." Industrial systems, which incorporate distributed manufacturing systems within them, cover the full spectrum of digital manufacturing, from peer-to-peer production through mass customisation, customised fabrication, mass fabrication, and personal fabrication.

The chapter starts off by providing a general introduction to AM technologies, including their characteristics and a general grasp of their industrial applications. The following section of the chapter delves deeply into some current technology examples that are being used and deployed by numerous manufacturing companies all around the world. These examples show how manufacturing facilities have already started implementing AM and the effects of doing so on the overall viability of the production system. These cases can be classified into four main categories by using a method based on product circuitry: product and process redesign, material input processing, manufacturing to sell components, and closing the loop.

3D printing, commonly referred to as "additive manufacturing," is a technique for producing 3D structures out of successive layers of material. It has the ability to completely transform manufacturing by making it possible to produce complex, customised parts on demand, cut down on waste and energy use, and employ new, environmentally friendly materials. To ensure that additive manufacturing is actually sustainable, as with any manufacturing process, it is crucial to take into account its effects on the environment and society. The use of sustainable materials is a crucial component of sustainability in additive manufacturing [3]. Conventional manufacturing methods frequently rely on non-renewable resources, such as fossil fuels and minerals, whose extraction and processing can have a severe impact on the environment. Contrarily, the use of a broad variety of materials is permitted by additive manufacturing, including recycled plastics, bioplastics manufactured from renewable resources, and even metal powders made from recycled metal. By utilising sustainable materials, additive manufacturing can minimise the environmental effects of material production and lower the demand for non-renewable resources.

One example of a sustainable material that has been used in additive manufacturing is polylactic acid (PLA), a bioplastic manufactured from corn flour or other renewable resources [4]. In comparison to conventional plastics, PLA has a number of benefits, including the ability to degrade naturally and a smaller carbon footprint. Additionally, it has a wide range of uses, including in consumer goods, automobile parts, and medical implants. It is crucial to remember that the sustainability of bioplastics depends on the precise feedstocks and manufacturing procedures employed and that they are not necessarily a more environmentally friendly choice than conventional plastics.

Recycled plastic is an additional promising material for additive manufacturing. Plastic recycling can lower the need for virgin plastic, which is frequently produced using non-renewable resources like oil or natural gas. Furthermore, recycling plastic helps cut down on the amount of trash dumped in landfills, where it can take hundreds of years for it to decompose [5]. The usefulness of recycled plastic for additive manufacturing, however, can vary depending on its quality and purity. Another potential sustainable additive manufacturing material is metal powder made from discarded metal. Recycling metal can lessen the need for mining, its negative effects on the environment, and the energy required for production [6]. Recycled metal powders may not be widely available, and their quality and purity may not always be consistent.

The amount of energy used in additive manufacturing methods is another crucial factor. Some 3D printing techniques can use a lot of energy, whereas others are often energy-efficient. In comparison to binder jetting and material jetting, laser sintering and material extrusion methods, for instance, often use less energy. Also, by removing the need for material transportation and minimising material waste, additive manufacturing can lower the energy usage of the entire manufacturing process.

For instance, cutting and shaping materials is a common part of traditional manufacturing processes, which can produce a lot of waste. Yet, since layers of material are only added where they are required, additive manufacturing can make things with little material waste. As a result, less material may need to be acquired, processed, and delivered, which could lead to a reduction in the overall amount of energy used.

By decreasing the requirement for transportation in addition to lowering material waste, additive manufacturing can further cut down on energy use and greenhouse gas emissions. Transporting raw materials, intermediate goods, and finished goods to and from different places is a common step in traditional manufacturing processes. As opposed to manufacturing methods, additive manufacturing enables the manufacture of parts as needed, possibly at or close to the point of use. This can reduce the need for transportation, as well as the energy use and emissions that go along with it.

8.2 HISTORY OF ADDITIVE MANUFACTURING

3D printing, also known as additive manufacturing, is the process of turning a digital model into a physical thing by adding layers to it. The first patent for 3D printing technology was submitted in the 1980s, which marks the beginning of additive manufacturing. The technology was not, however, widely used or accessible until the 21st century. Chuck Hull, who created a method called stereolithography, submitted the first patent application for a 3D printing procedure in 1986. In this method, a photosensitive resin is exposed to UV light under controlled conditions, hardening the material to produce a solid object [7].

Other 3D printing technologies were created over the ensuing decades, such as selective laser sintering and fused deposition modelling, which both use nozzles to extrude molten material layer by layer to build objects. Selective laser sintering uses a laser to sinter (or fuse) powders or other materials to create solid objects.

The use of 3D printing technology expanded in the 21st century as it became more generally accessible and usable for a range of purposes, including prototypes, production, and domestic use [8]. In a number of sectors, including aerospace, automotive, healthcare, and architecture, additive manufacturing has seen significant expansion in recent years.

As 3D printing technology continues to advance and becomes more widely available, it is expected to play an increasingly important role in the future of manufacturing and product development.

8.3 ROLE OF SUSTAINABILITY IN ADDITIVE MANUFACTURING

Because less material is used in the production of finished goods, AM has a variety of benefits in terms of the environmental impact of the conversion of fresh materials into finished and processed goods. These benefits also improve the capacity of the components of intricate geometrical designs to be light in weight. AM offers much better operational, financial, and environmental results. The biggest benefit of AM is how inexpensively it can finish conventional particle synthesis in small state manufacturing batches [9]. Since additive manufacturing (AM) relies on peer-to-peer material requirements, the biggest barrier to manufacturing any product is the cost of materials.

8.4 SCULPTURES: THE SOLUTION OF SUSTAINABILITY IN ADDITIVE MANUFACTURING

AM has shown benefits in many interdisciplinary applications. Some of these applications are unworthy and difficult to explain, while others remain unknown. When it comes to applications of additive manufacturing in sculpture construction, it has been done to create potential responses of original artefacts so that original sculptures can be replaced from display and preserved in storage spaces [10]. There are also some replicas of such sculptures that have been installed with other original sculptures in places like the famous Forbidden City in Beijing, China. What comes to mind as an underlying observation is that the potential use of AM in the construction of complex structures has already been exploited by fine ants as well. The intersection of knowledge in traditional arts, such as sculpture, and digital technologies has proven to be a cost-effective and optimal method for preserving cultural pieces of history in the most appropriate ways. There is a direct void when it comes to the practise and knowledge of such interdisciplinary techniques combined with the complex construction machinery in use as compared to other engineering and medical applications. Some active industry involvement is required to improve the optimal digital interface, streamline the processing chains, uphold these industries, and provide meaningful alternative solutions to the sculpture maker community. If the above is done, it will result in the optimisation of complex industrial procedures and a reduction in the time consumption of TDS and reputational tasks. Knowing that art and culture are components that are meaningful to sculptors and subjective to aesthetes will help you understand anything in this. There are several case studies of the previously discussed process of sculpture creation through AM, which has played a significant and meaningful role in the sculpture industry. Two important cases are the horases project carried out in Manalisa and the renovation project of the Forbidden City, where some statues of the city that were damaged during the excavation had their specific model parts recreated to match their complexity, ancient look, and durable nature.

There are still many shortcomings in making sculptures or statues through AM that have to be investigated further for better exposure of the proposed technology in the field of sculpture manufacturing:

1. There are still many objects that can be scanned easily; such objects comprise amber, jade, pearls, etc. because of their translucent nature, i.e., material objects that have very high reflective surfaces or very low reflective surfaces (e.g., silver, platinum, gold, and hardwood).
2. If the scanning process is being carried out in a confined space, then detailed mapping of an object is hard to configure.
3. A fine level of finishing grade is difficult to achieve with cost-effective materials in the sculptor's AM process [11].

The long-term impact of AM techniques on the traditional sculpture manufacturing industry can be understood by comparing it to the impact that cameras and photography had on the painting industry after their invention and widespread general use. It is interesting to note that both of them are still coexisting simultaneously, despite the fact that painting should have become obsolete after the invention of the camera, but what was predicted didn't actually happen. The art industry will continue to exist in the modern era, when various AI platforms can generate art that mimics the human mindset. It is important to understand that modern technology will be absorbed by the artistic community in specific ways.

8.4.1 Using sustainable material in sculpting

The inception of 3D printing led to a rise in demand for different materials, such as plastic, fibres, carbon nanotubes, ceramics, metals, and composites, to make parts with the technology. It is very important to pick up the right material for the right purpose and process. Thus, the process of picking the right materials encompasses many factors such as texture, cost, material technology, printing technology, high-end demands, and complexity of the model.

Some materials are widely used in AM to meet the demands of the end-user product. Nylon, or polyamide, which is a thermoplastic material that is synthetic in nature and generally known as plastic, is commonly used in selective laser sintering (SLS) [12].

Stainless steel is durable, ductile, strong, and corrosion-resistant due to its nature. It is widely used in various assistive technologies that supplement AM. But as 3D printing of stainless steel is not only hard and time-consuming but also a costly affair, it allows further investigation of materials that have the same advantages as stainless steel but can cover up its disadvantages [13]. Hence, if we use slag as a replacement in the manufacturing of certain products that require the same material characteristics as stainless steel, then a cost-effective way can be adopted to replace steel in AM with a better alternative. The process is to be carried out through direct metal laser sintering, i.e., also used to print stainless steel [13].

Stainless steel slag can be taken up as a potential material that has a wide array of possibilities, i.e., is strong, heat-resistant, lightweight, and chemically resistant. Due to its strenuous nature, it is an incredibly optimal material to be maintained by laser sintering tools in AM, which makes it an appropriate material for additive manufacturing [14].

8.5 TERRACOTTA SLAG – A NEW ADDITIVE MANUFACTURING MATERIAL

It is possible to create a new material by mixing terracotta clay and slag, and this material can have unique properties and characteristics depending on the specific proportions of clay and slag used in the mix. Terracotta clay is a type of clay that is commonly used in the production of ceramics, and it is known for its ability to be moulded and shaped into a variety of forms. Slag, on the other hand, is a by-product of the process of firing terracotta clay, and it is composed of various materials, including clay, sand, and other minerals.

When mixed together (Figure 8.1), terracotta clay and slag can create a material with improved strength and durability, as the slag can help to increase the density and reduce the porosity of the material. The material may also have improved firing properties, making it easier to shape and form into a desired shape [15]. The specific properties of the new material will depend on the proportions of clay and slag used in the mix. A higher percentage of slag in the mix may result in a stronger and more durable material, but it may also make the material more difficult to work with. The optimal ratio of slag to clay will depend on the specific requirements of the application. The addition of slag to terracotta clay can improve the properties of the finished product in several ways. For example, slag can help to improve the strength and durability of the ceramics, as it can increase the density and reduce the porosity of the material. Slag can also improve the firing properties of the clay, making it easier to shape and form into a desired shape. The amount of slag that is mixed with terracotta clay can vary depending on the desired properties of the finished product [16]. Generally, a higher percentage of slag in the mix will result in a stronger and more durable product, but it may also make the clay more difficult to work with. The optimal ratio of slag to clay will depend on the specific requirements of the application.

8.5.1 Extraction of slag

To understand how slag is gained, we first need to understand the process of converting iron to steel. Initially, iron ore feedstock is placed in a primary steel-making furnace, where a blower is used to force oxygen through molten iron, resulting in a reduction of carbon content while simultaneously removing impurities from the leftover material, known as furnace slag. The material recovered after the process is forwarded to a furnace known as

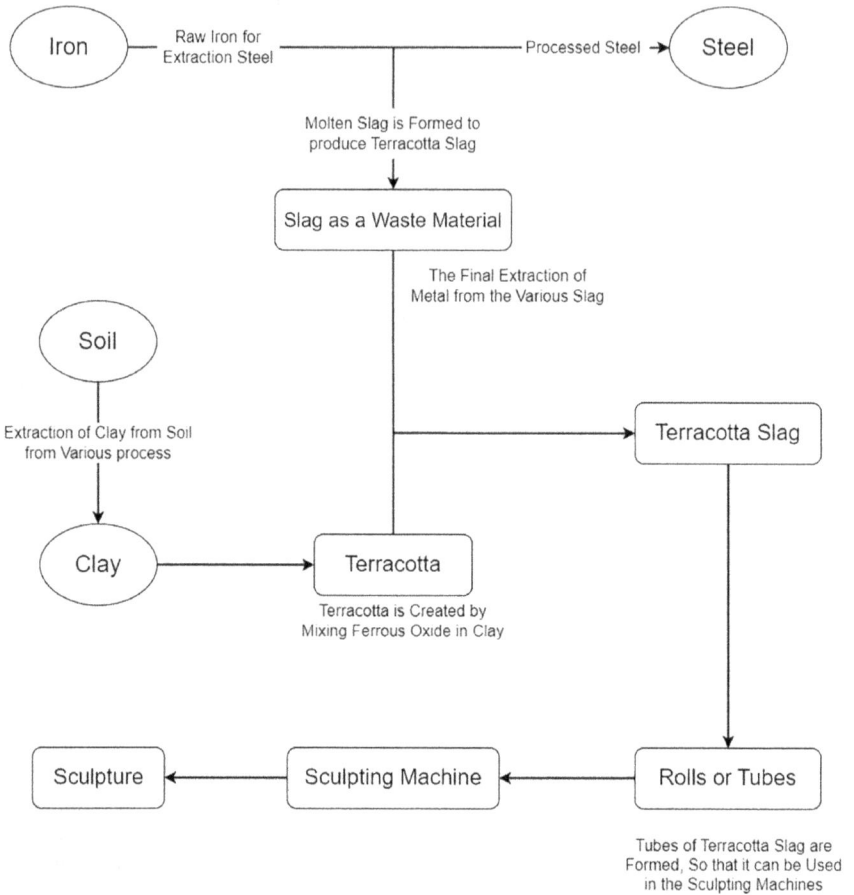

Figure 8.1 Flowchart for making terracotta slag.

a transfer ladle, where the molten material, in this case metal from iron ore, is poured in large amphorae. After some time, the ladle slag comes out as waste material, and the product is forwarded to the ladle waste and dumped residue chamber, where the cleanout or pit slag is received as the waste product [17].

Now we receive furnace slag, raker slag, ladle slag, cleanout slag, or pit slag from the metal recovery procedures carried out in different stages to recover the lost material in the form of waste material to check if such material can be further utilised for meaningful purposes (Figure 8.2). The slag residue retrieved at different stages of the process can be differentiated into metallic and non-metallic substances, which have to be again extracted (Figure 8.3). The non-metallic steel slag can be utilised for potential aggregate use, while the metal slag that we receive after further extraction of materials through the metal recovery process could be sent to the construction industry and

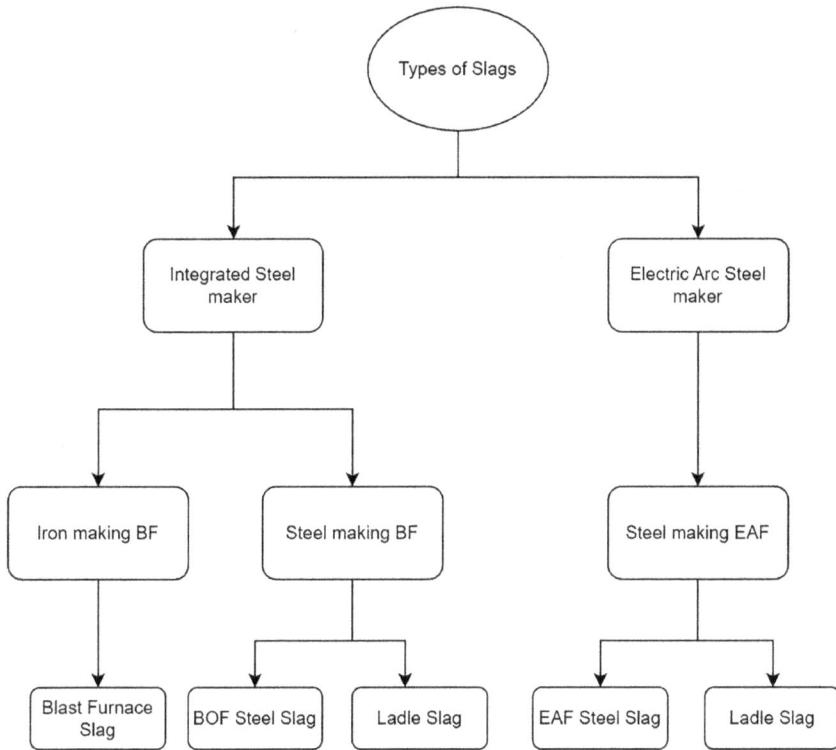

Figure 8.2 Types of slag produced during the production of steel [22].

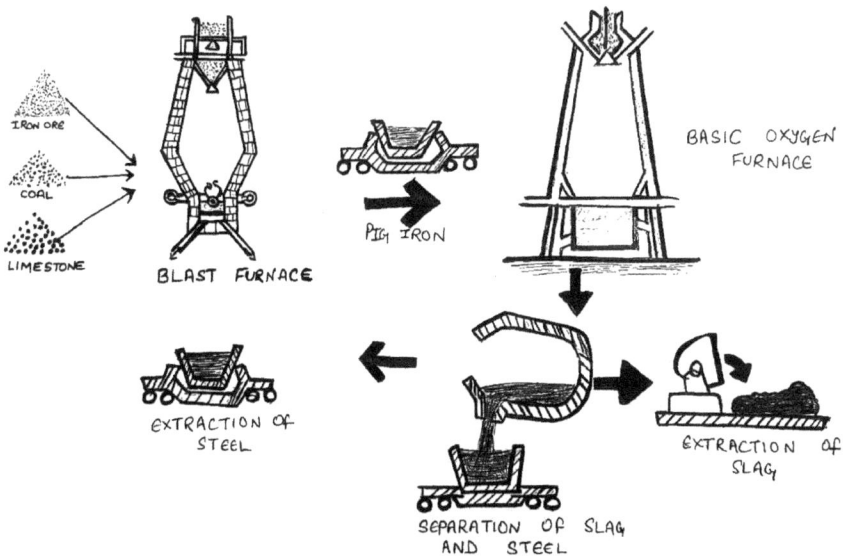

Figure 8.3 Process of producing different types of furnace slag [22].

roads department, where it can be mixed with tar, i.e., the essential road construction material, with which we can achieve production-grade road construction material that will have longer durability. It can also be mixed with cement to improve its tenacious nature in holding up to or withstanding the structural complexity of the building [17].

The above examples are exemplary enough to try various domains where slag can be infused to be used as production-grade material, and its utilisation with AM technologies is something that has not been tried before, thereby providing a domain worth exploring for research and development. The slag that is received after the differentiating mash to obtain metallic and non-metallic slag, these are both production-grade materials. That is because the compounds are predominantly aromatic.

Present in the slag are calcium-magnesium iron oxide, calcium-magnesium iron dioxide, dicalcium silicate, tricalcium silicate, dicalcium ferrite, calcium aluminate, merwinite, some free lime, and some free magnesia. The proportion in which these compounds are retrieved, on the other hand (Table 8.1), is subject to the process used in the steel-making process and practise, or it may also depend on the cooling rate of steel slag. There are some mechanical properties that are subject to the processed steel slag, i.e., the abrasion resistance of the slag is good, and it has a high bearing strength. As the said material has a high heat capacity, aggregates of steel slag can hold heat for a longer duration in comparison to their natural aggregates.

Slag is a by-product of the smelting process used to produce iron. It is formed when the impurities in iron ore are smelted by a various procedure to produce iron, and as a by-product, we get slag. It is formed when impurities in the iron ore are separated out and removed during the smelting process. In this process, the iron ore is heated to a high temperature in a furnace along with a reducing agent, such as coke or charcoal. The reducing agent reacts with the impurities in the ore, separating them out as slag. The slag is then removed from the furnace and allowed to cool. There are

Table 8.1 Flowchart of making terracotta slag [21]

Constituent	Composition (%)
CaO	40–52
SiO$_2$	10–19
FeO	10–40
	(70%–80% FeO, 20%–30% Fe$_2$O$_3$)
MnO	5–8
MgO	5–10
Al$_2$O$_3$	1–3
P$_2$O$_5$	0.5–1
S	<0.1
Metallic Fe	0.5–10

several different techniques that can be used to produce slag from iron ore. One common method is the blast furnace process, which has been used for centuries to produce iron. In this process, the iron ore is mixed with coke, a form of coal, and limestone. The mixture is then fed into the top of a blast furnace, where it is heated to a high temperature.

As the mixture moves through the furnace, the coke reacts with the oxygen in the air to produce carbon dioxide. The carbon dioxide reacts with the iron ore, reducing it to molten iron. The impurities in the ore, such as silica, alumina, and phosphorus, are separated out as slag. The slag is lighter than the molten iron, so it floats to the top of the furnace and is removed through a slag hole.

Another method for producing slag from iron ore is the direct reduction process (Figure 8.4). In this process, the iron ore is mixed with a reducing agent, such as natural gas or hydrogen, and heated to a high temperature. The reducing agent reacts with the impurities in the ore, separating them out as slag. The slag is then removed from the furnace and allowed to cool. The specific process used to produce slag from iron ore will depend on the type of iron ore and the desired end product. Some common techniques used in iron ore smelting include the blast furnace process and the direct reduction process [18]. In addition to being a by-product of the iron smelting process, slag has a number of important uses. It can be used as a construction material, particularly in road construction, and as a raw material in the production of cement. It can also be used as an abrasive material due to its hardness and toughness. Slag can also be used as fertiliser, as it contains nutrients that are beneficial to plants. Overall, the production of slag from iron ore is an important process that helps remove impurities from the ore and produce a high-quality product. It is a complex process that involves several different techniques, depending on the type of iron ore and the desired end product.

8.5.2 Mixing of terracotta clay with slag

Iron slag can potentially be mixed with clay to form sculptors, but there are a few factors to consider before doing so. One of the main properties of iron slag that may be of concern when using it in sculpture is its high melting point. Iron slag has a melting point of around 1,500–1,600°C, which is significantly higher than the melting point of most clays. This means that the iron slag may not fully integrate with the clay and may remain as a separate phase in the finished sculpture.

Another factor to consider is the chemical composition of the iron slag. Iron slag typically contains a high amount of silica, which can be reactive with certain types of clay. Depending on the specific clay and iron slag used, the combination of the two materials may result in unwanted chemical reactions that could affect the properties of the final sculpture. It is also worth noting that iron slag is typically very hard and dense, which may

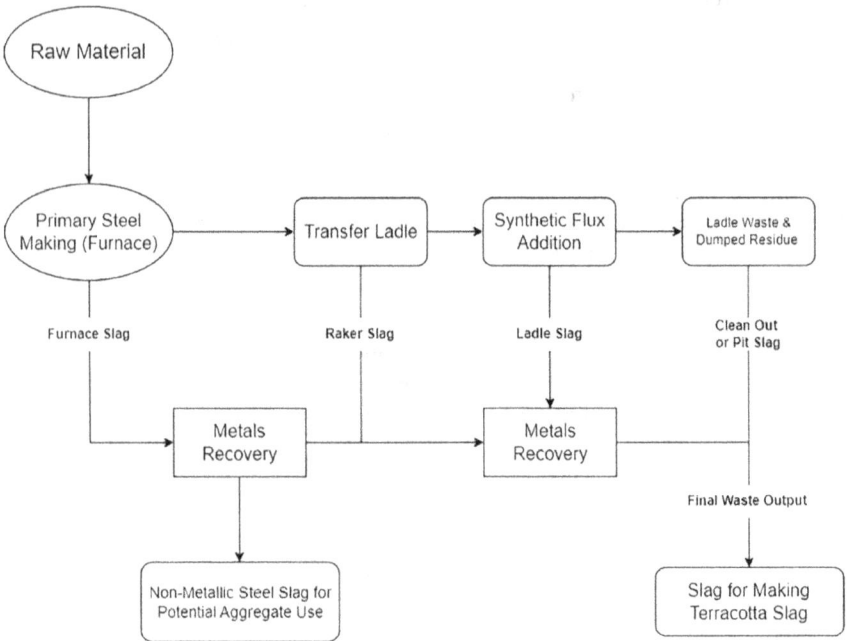

Figure 8.4 Flowchart for making slag [21].

make it difficult to work with when sculpting. It may require special tools and techniques to shape and sculpt the material, and it may not have the same flexibility and malleability as clay.

Overall, while it is possible to mix iron slag with clay to form sculptures, there are several factors that need to be taken into consideration before attempting to do so. It may be necessary to perform tests and experiments to determine the most suitable combination of materials and techniques for your specific project. Mixing terracotta clay with metal slag can create a new material with properties that are intermediate between those of the individual components. The strength, durability, and other characteristics of the mixture will depend on the specific types of terracotta clay and metal slag used, as well as the proportions of each material. To understand the mixing of terracotta clay and metal slag, it is helpful to first understand the properties of these materials individually [19].

Terracotta clay is a naturally occurring material made up of fine-grained minerals, typically aluminium silicates. When mixed with water and formed into a desired shape, terracotta clay can be fired at a high temperature to create a durable, hard material. The properties of terracotta clay can vary depending on the specific minerals it contains and the conditions under which it was formed. When mixed together, terracotta clay and metal slag can create a material with properties that are intermediate between those of the individual components. The strength and durability of the mixture will depend on the specific types of terracotta clay and metal slag used, as well as the proportions of each material. The resulting mixture may also have other properties that are influenced by the specific characteristics of the individual components.

There are several factors that can affect the properties of the terracotta clay and metal slag mixture, including the particle size of the individual components, the proportions of each material, and the specific types of terracotta clay and metal slag used. For example, using fine-grained terracotta clay and metal slag with smaller particle sizes may result in a mixture with higher strength and greater durability than one made with coarser materials. Similarly, using a metal slag with a higher silica content may result in a mixture with greater strength and durability than a mixture made with a metal slag with a lower silica content.

In general, the mixing of terracotta clay and metal slag can result in a material with a range of possible properties, depending on the specific characteristics of the individual components and the proportions in which they are mixed. It is important to carefully consider the specific properties and characteristics of the materials you are using before mixing them together and to carefully control the proportions of each material to achieve the desired properties in the resulting mixture.

8.6 ADVANTAGES OF MIXING TERRACOTTA CLAY AND SLAG

The potential advantages of mixing terracotta clay and slag may depend on the specific types of terracotta clay and slag used, as well as the proportions of each material.

Improved strength and durability: Depending on the specific characteristics of the terracotta clay and slag used, mixing these materials would result in a mixture with improved strength and durability compared to either material on its own.

Enhanced properties: The specific properties of the terracotta clay and slag mixture may be influenced by the characteristics of the individual components. For example, using a metal slag with a high silica content may result in a mixture with improved insulation properties, while using a metal slag with a high oxide content may result in a mixture with improved abrasive properties.

Cost savings: Mixing terracotta clay and slag could potentially result in cost savings compared to using either material on its own. For example, using slag as a partial replacement for terracotta clay could potentially reduce the overall cost of the mixture.

Waste reduction: Mixing terracotta clay and slag can also help reduce waste by repurposing materials that would otherwise be discarded [20].

It is important to note that the specific advantages of mixing terracotta clay and slag will depend on the specific characteristics of the individual components and the proportions in which they are mixed. It is advisable to carefully consider the properties and characteristics of the materials you are using before mixing them together and to carefully control the proportions of each material to achieve the desired properties in the resulting mixture.

8.7 DISADVANTAGES OF MIXING TERRACOTTA CLAY AND SLAG

Reduced strength and durability: Depending on the specific types of terracotta clay and slag used and the proportions of each material, the resulting mixture may be less strong or durable than either material on its own.

Changes in other properties: Mixing terracotta clay and slag may also result in changes in other properties of the mixture, such as shrinkage or thermal expansion, that could affect its performance or suitability for certain applications.

Compatibility issues: The compatibility of terracotta clay and slag may depend on the specific types of each material used. In some cases, mixing these materials may result in chemical reactions or other interactions that could affect the properties of the mixture [20].

Quality control challenges: Mixing terracotta clay and slag may also present challenges in terms of quality control, as it may be difficult to consistently achieve the desired proportions of each material and to predict the resulting properties of the mixture.

It is important to carefully consider the specific disadvantages of mixing terracotta clay and slag before using this material in any application. It may be advisable to test the properties of the mixture in small-scale experiments before using it on a larger scale.

8.8 CONCLUSION

Additive manufacturing, also known as 3D printing, is the process of creating a physical object by building it up layer by layer from a digital model. It has the potential to be a more sustainable manufacturing process compared to traditional methods, as it can reduce waste and energy consumption and enable the production of customised and complex products with minimal material use. One potential benefit of additive manufacturing is that it can reduce the amount of material waste produced during the manufacturing process. Traditional manufacturing methods often involve cutting or shaping materials, which can generate significant amounts of waste. In contrast, additive manufacturing builds up an object layer by layer, allowing for the use of only the material that is needed to create the final product. This can lead to significant material savings and reduced waste generation. Another potential benefit of additive manufacturing is that it can enable the production of customised and complex products with minimal material use. Traditional manufacturing processes often require the production of large quantities of a single product, which can result in excess material use and waste. In contrast, additive manufacturing allows for the production of customised products on demand, which can reduce the need for excess material and waste. However, there are also some challenges to the sustainability of additive manufacturing. One challenge is that the energy efficiency of 3D printers can vary significantly depending on the type of printer and the material being used. Some 3D printers are more energy-efficient than others, and the energy required to produce a product can depend on the complexity of the product and the type of material being used. Another challenge is that the sustainability of additive manufacturing can depend on the type of material being used. Some materials, such as plastics, may not be as sustainable as others, particularly if they are not recycled or disposed of properly. It is important to consider the environmental impacts of the materials used in additive manufacturing in order to assess the overall sustainability of the process. Overall, the sustainability of additive manufacturing depends on a variety of factors, and it is difficult to draw a general conclusion about its sustainability. It is important to consider the specific context and circumstances in which additive manufacturing is being used in order to assess its sustainability.

REFERENCES

[1] Mehrpouya, M., Vosooghnia, A., Dehghanghadikolaei, A., & Fotovvati, B. (2021). The benefits of additive manufacturing for sustainable design and production. In *Sustainable Manufacturing* (pp. 29-59). Elsevier, ISBN 9780128181157.
[2] Grosse-Sommer, A. P., Grünenwald, T. H., Paczkowski, N. S., Van Gelder, R. N. M., & Saling, P. R. (2018). Sustainability strategy. CDS/S - C 104, 67056, Ludwigshafen, Germany: BASF SE.
[3] Farina, I., Singh, R., Kumar, R., & Colangelo, F. (2019). Multi-material additive manufacturing of sustainable innovative materials and structures. *Polymers*, 11(1), 62. https://doi.org/10.3390/polym11010062.
[4] Nagarajan, V., Mohanty, A. K., & Misra, M. (2016). Perspective on polylactic acid (PLA) based sustainable materials for durable applications: focus on toughness and heat resistance. *ACS Sustainable Chemistry & Engineering*, 4(6), 2899–2916. https://doi.org/10.1021/acssuschemeng.6b00321.
[5] Al -Salem, S. M., Lettieri, P., & Baeyens, J. (2009). Recycling and recovery routes of plastic solid waste (PSW): a review. *Waste Management*, 29(10), 2625-2643. https://doi.org/10.1016/j.wasman.2009.06.004.
[6] Damgaard, A., Larsen, A. W., & Christensen, T. H. (2009). Recycling of metals: accounting of greenhouse gases and global warming contributions. *Waste Management Research*, 27(8), 773–780. https://doi.org/10.1177/0734242x09346838.

[7] Chae, H., Huang, E. W., Woo, W., Kang, S. H., Jain, J., An, K., & Lee, S. Y. (2021). Unravelling thermal history during additive manufacturing of martensitic stainless steel. *Journal of Alloys and Compounds*, 857, 157555. https://doi.org/10.1016/j.jallcom.2020.157555.

[8] Kodama, H. (1981). Automatic method for fabricating a three-dimensional plastic model with photo-hardening polymer. *Review of Scientific Instruments*, 52(11), 1770–1773.

[9] Kellens, K., Mertens, R., Paraskevas, D., Dewulf, W., & Duflou, J. R. (n.d.). Environmental impact of additive manufacturing processes: does AM contribute to a more sustainable way of part manufacturing? *Procedia CIRP*, 29, 39–44. https://doi.org/10.1016/j.procir.2015.01.039.

[10] Elsen, A. E. (2003). *Rodin's Art: The Rodin Collection of the Iris & Gerald B. Cantor Center for the Visual Arts*. Oxford, United Kingdom, Oxford University Press. ISBN 0-19-513381-1

[11] Rajaguru, K., Karthikeyan, T., & Vijayan, V. (2017). Additive manufacturing - state of art. *Materials Today: Proceedings*, 4(2), 3178–3185. https://doi.org/10.1016/j.matpr.2017.02.279.

[12] Gan, X., Fei, G., Wang, J., Wang, Z., Lavorgna, M., & Xia, H. (2020). Powder quality and electrical conductivity of selective laser sintered polymer composite components. In *Structure and Properties of Additive Manufactured Polymer Components* (pp. 149-185). Sawston, United Kingdom: Woodhead Publishing.

[13] Irawan, A., Zadi-Maad, A., & Rohib, R. (2018). The use of additive manufacturing in the production of aerospace components: a review. *IOP Conference Series: Materials Science and Engineering*, 285(1), 012028. https://doi.org/10.1088/1757-899X/285/1/012028

[14] de Brito, J., & Agrela, F., eds. (2018). *New Trends in Eco-efficient and Recycled Concrete*. Sawston, United Kingdom: Woodhead Publishing.

[15] Britannica, T. Editors of Encyclopaedia. (2023, June 2). *Terra-cotta*. Encyclopedia Britannica.

[16] Singer, F., & Singer, S. S. (1971). *Industrial Ceramics*. United Kingdom: Chapman & Hall.

[17] Fruehan, R. (1998). *The Making, Shaping, and Treating of Steel, Steelmaking and Refining Volume*, 11th Edition. Pittsburgh: The AISE Steel Foundation.

[18] Cwirzen, A. (2020). *Properties of SCC with Industrial By-products as Aggregates*. doi: 10.1016/B978-0-12-817369-5.00010-6.

[19] Ortega-López, V., Manso, J. M., Cuesta, I. I., González, & J. J. (2014). The long-term accelerated expansion of various ladle-furnace basic slags and their soil-stabilization applications. *Construction and Building Materials*, 68, 455–464.

[20] Vranić, A., Bogojevic, N., Ciric-Kostic, S., & Croccolo, D. (2017). Advantages and drawbacks of additive manufacturing. *Facta Universitatis, Series: Mechanical Engineering,* 15(2), 57–70. https://doi.org/10.5937/IMK1702057V.

[21] US Department of Transportation. (1997). User guidelines for waste and byproduct materials in pavement construction (Publication Number: FHWA-RD-97-148).

[22] Tiwari, M. K., Bajpai, S., & Dewangan, U. (2016). Steel slag utilization - Overview in Indian perspective. *International Journal of Advanced Research*, 4(11), 1110–1115. https://doi.org/10.21474/IJAR01/1442.

Chapter 9

Towards a greener future
How additive manufacturing and bio-based materials are saving the environment

Jagdeep Kaur, Sahil Srivastava,
Anil Dhanola, and Sachin Moond
Chandigarh University Gharuan

9.1 INTRODUCTION

Additive manufacturing (AM) is leading to a future where many applications were previously carried out from a tentative location, i.e., where manufacturing of such utilities was possible, but nowadays one can print a replacement part for their car from the comfort of their own garage or create a custom-fit prosthetic, be it an arm or leg, for a loved one using nothing but a computer and a 3D printer, with the comfort of building it from their home. This is the promise of AM, a technology that is rapidly changing the way we think about production, and consumption. AM, commonly also known as 3D printing, is a type of manufacturing, production and fabrication technology that creates three-dimensional objects by layering materials of different attributes, texture properties, and complex natures based on a computational model in 3D design software. It is a versatile technique that can be utilised to enhance and produce a wide range of products, including parts with complex geometries and one-of-a-kind designs. During AM, a digital representation of the object is created employing computer-aided design (CAD) software. This design is then separated into thin layers and placed in a 3D printer, which uses a variety of techniques to deposit or cure material layer by layer to produce the item. A variety of materials, including plastics, metals, ceramics, and even living cells, may be used.

One of the key benefits of AM is its ability to create customised objects, products, and utilities on demand at par with the customer experience, which can be especially useful in fields such as medical, aerospace, construction, automotive, consumer product utilities, foods, education, and the bio-based product industry. For example, medical professionals can use 3D printing to create custom-fit prosthetics or implants that are tailored to the specific needs of individual patients. Similarly, aerospace companies can use 3D printing to produce custom-fit parts for their aircraft, which can reduce weight and improve performance. In addition to custom manufacturing,

DOI: 10.1201/9781003406488-9

AM can also be used to produce small, medium, and large quantities of identical or customised objects quickly and efficiently. This can be especially useful for small businesses or organisations that need to produce a large number of items but may not have the resources or equipment to do so using traditional manufacturing methods.

While AM has the potential to bring many benefits to society, it also raises a number of challenges and questions. One concern is the potential for job loss as traditional manufacturing processes are replaced by 3D printing. Another concern is the environmental impact of AM, as it may rely on non-renewable energy sources and produce waste materials. Despite these challenges, the potential for AM to revolutionise the way we produce and consume goods is undeniable. In this chapter, we will delve into the exciting world of 3D printing and explore its current and potential future applications in fields such as medicine, aerospace, and manufacturing. We will also consider the challenges and ethical questions raised by this technology as it continues to evolve and make an impact on society. In recent years, there has been an increase in the use of bio-based materials in AM for several reasons. Initially, bio-based materials are often more sustainable and environmentally friendly than traditional materials that have been put to use, such as plastics, which are made from non-renewable fossil fuels and cause an excess of pollution. The use of bio-based materials in AM can help reduce the carbon footprint of manufacturing processes and contribute to a more sustainable future, which will eventually lead to the betterment of society. Secondly, bio-based materials have been used to create a variety of products with unique properties, textures, attributes, and characteristics. For example, wood-based materials can be used to generate products with a wood-like appearance, while plant-based materials can be used to create products with a natural, organic look and feel. Third, bio-based materials can be used to create products that are biocompatible with the human body, and they can also be used to replace materials with harmful effects, such as plastic, that are used in large-scale manufacturing operations. For example, bio-glass and other bio-based materials can be used to create medical implants and drug delivery systems that are compatible with the human body and can help reduce the risk of rejection or infection. Such materials can also be used to make toys for kids that are eventually made from harmful elements, which are dangerous as such materials can cause hindrance in their eventual growth during their growing years. It might also be beneficial in the long run, as there are biomaterials that are germ resistant. Finally, the increasing demand for sustainable and environmentally friendly products has driven the development and use of bio-based materials in AM. As consumers become more aware of the environmental impact of the products they purchase, companies are seeking out sustainable materials and production methods to meet this demand.

In this chapter, we will delve into the exciting world of 3D printing and examine the ways in which it is transforming industries and improving lives, and we will also look into how bio-based materials such as hemp and mycelium-based bio-composites can be used for making bedsheets, clothes, aprons, curtains, etc. with hemp, which is a thermally resistant material and thus can be useful in preventing big damages caused by fire. We will also look at how the waste produced from such AM processes can be used for making fertiliser, which can nullify waste generation in all likelihood.

9.2 INDUSTRIES WHEREIN AM CAN BE EMPLOYED TO BENEFIT SOCIETY

1. *Medicine:* AM can be used to create customised medical devices that were previously made using traditional processes, resulting in an increase in the cost of medical applications, preventing the middle class from benefiting from such Technological advancements such as prosthetics, implants, and surgical instruments for critical parts of the body, and thus the cost optimisation of such practises, will lead to the global improvement of society. It may also be used to create custom-fit drug doses that are adjusted to the exact needs and features of the individual patient, as shown in Figure 9.1. AM can be used to create custom-fit medication dosages by using 3D printing to produce pre-scribed dosage forms such as pills or capsules on an individual level for patients rather than producing and providing such medicines on an industrial scale, resulting in increased efficiency of such medical utilities [1]. This enables precise management of the medicine dose, ensuring that the patient receives the optimum quantity of medication to produce the intended therapeutic benefit while having no unwanted effects. Another inventive way that AM can be used in the production of custom-fit medication dosages is to use 3D printing to create customised drug delivery devices such as patches or inhalers from bio-based germ-resistant materials, which will also prevent the inhibition of germ extracts on the surface of such utilities [2].

2. *Aerospace:* AM has the ability to transform the aerospace and aeronautical industries by enabling the manufacturing of complex geometrical and customised parts and components in less time and at a lower cost. AM is being utilised or has the potential to be used in the aerospace sector in the following ways: *Production of aircraft parts:* AM may be used to create complicated and personalised aviation parts, including engines, fuel nozzles, and other components. This may result in time and cost savings as compared to traditional manufacturing processes, as AM allows for the manufacture of components on-demand and removes the need for tooling [3].

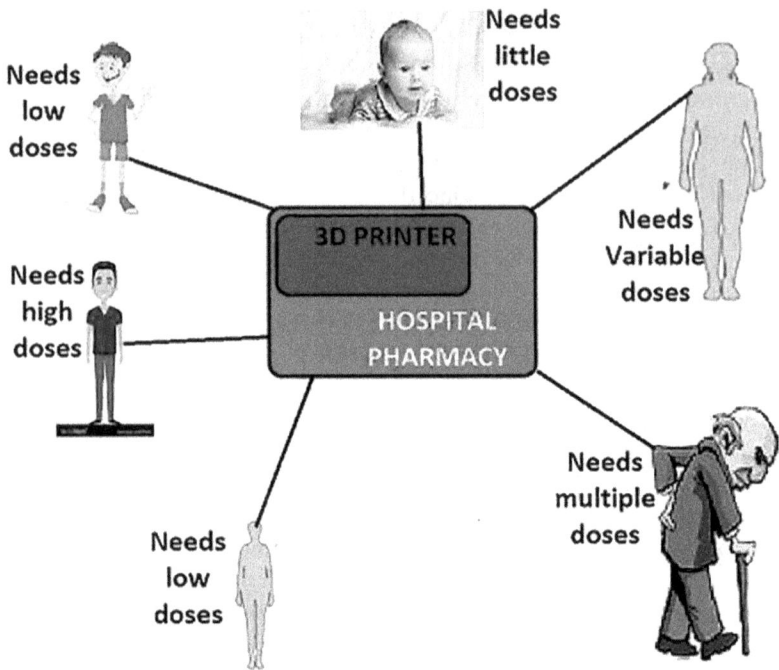

Figure 9.1 Based on the diagnosis, treatment, and treatment plan.

Repair and maintenance of aircraft parts: AM may be used to repair and maintain aircraft components by manufacturing replacement parts on demand. This can save on downtime and the expenses of locating and replacing parts. *Production of satellite components*: AM may be used to fabricate complicated and specialised satellite components such as antennas, solar panels, and other components. When compared to standard production processes, this can shorten lead times and save costs [4]. *Production of space vehicles:* AM may be used to create sophisticated and personalised space vehicles such as spacecraft and rovers. This enables the construction of tailored and optimised vehicles for individual missions, lowering costs and lead times [5]. Overall, the application of AM in the aerospace sector has the potential to enhance productivity, cut costs, and enable the creation of customised and complicated parts and components [6].

3. *Construction:* Additive printing may be used to make custom-fit construction components such as wall panels and structural beams. It may also be used to produce economical housing options in locations

with limited access to typical building materials. It also permits the development of complicated architectural geometries, which were previously used to cost a tonne of money with little customer pleasure because the finishing of such architectural marvels was not close to the intended outcomes [7]. As a result, there was no practical innovation in the creation of such architectural wonders, but with the assistance of AM, such architectural marvels may now be constructed from unfriendly building materials, increasing the lifetime of such projects [8].

4. *Automotive:* The automobile sector has the potential to be transformed by AM. AM is being utilised or has the potential to be used in the automobile industry in the following ways:

   Repair and maintenance of car parts: AM may be used to repair and maintain automobile parts by generating replacement parts on-demand. This can save downtime and the expenses involved with locating and replacing replacement components. *Production of custom car parts:* AM may be used to create customised automotive parts such as dashboard panels, door handles, and other components. This enables the manufacturing of customised and personalised automobiles, thus enhancing consumer satisfaction and sales.

   Production of prototype car parts: AM may be used to swiftly and cost-effectively build prototypes of automobile components, allowing for rapid prototyping and testing of new designs, resulting in cost optimisation of such measures, making manufacturing and development of newer designs more practical [9].

5. **Consumer products**: AM can be used to produce customised products such as jewellery, phone cases, and home decor items. The items may be manufactured with personalised choices, allowing them to blend in with the standard style and appearance of the house [10].

6. *Food*: AM may be used to create custom-fit nutrition bars and other food items. It may also be used to make edible culinary items out of a variety of components, such as chocolate and sugar [11]. Table 9.1 briefly highlights the firms as well as the technology that they have used to print specific food goods with the models of their printers that became available in that particular year. Previously, the technique was used to create edible plates and utilities that may be used as eatables after the meal on the cutlery has been consumed. Such cleverness is only conceivable when AM is accessible since there is no strong demand for such items at a certain penetration point; hence, bulk production of such utilities will result in further waste creation.

7. *Education*: In educational contexts, AM may be used to teach students about design, engineering, and manufacturing fundamentals. It may also be used to produce tailor-made teaching materials and tools

Table 9.1 Different technologies adopted by different companies

Technology	Company	Year	Materials
Material extrusion	PancakeBot	2010	Pancake
Material extrusion	Zmorph	2013	Chocolate and cake butter
Material extrusion	Natural Machines	2015	Multi-material Foodini
Material extrusion	Beehex	2016	Pizza, cake decorations
Material extrusion (melting capable)	ChocEdge	2010	Chocolate
Material extrusion (melting capable)	byFlow	2010	Chocolate
Material extrusion (melting capable)	Essential Dynamics	2012	Chocolate and others
Material extrusion (melting capable)	Structur3d	2013	Chocolate, frosting
Material extrusion (melting capable)	ORD solutions	2015	Chocolate, pasta
Material extrusion (melting capable)	Prosusini	2015	Chocolate, icing, cream cheese, jams, mixes, batters
Material extrusion (melting capable)	Becoda	2015	Chocolate
Material extrusion (melting capable)	Shiyin Tech	2016	Chocolate, cheese, jam, candy, biscuit, multi-material
Binder jetting	Dovetailed	2014	Fruit
Binder jetting	FoodJet printing systems	2015	Chocolate, liquid dough, sugar icing etc.

that will improve the efficiency of daily teaching activities. It will also be highly useful if this technology is made available to students for prototyping, since it will open up a wide range of academic options.

9.3 ROLE OF ADDITIVE MANUFACTURING IN BETTERMENT OF SOCIETY

There are several ways that AM may assist society, such as by boosting sustainability. AM can help decrease waste material and resource consumption since it enables the manufacturing of customised and on-demand items in real time without requiring vast amounts of material or energy. This can assist in minimising manufacturing's environmental impact and promoting sustainability. For example, AM can be used to produce customised and on-demand products such as spare parts and tools, reducing the need for large stockpiles of these items and reducing waste. Additionally, there are products or parts that are no longer made, i.e., due to regularised component upgrades on a daily scale, so when restoration of such products is required to increase their lifespan, finding such parts proves to be a hassle, so relying on AM could be greatly beneficial in that AM may also be used to repair and maintain current items so prolonging their lifespan and lowering the need for replacements [12]. In the case of a crisis, AM may also improve disaster response by allowing for the speedy and cost-effective manufacturing of required supplies and equipment. Due to destroyed infrastructure or other logistical issues, it may be difficult to get the required supplies and equipment in the aftermath of a disaster. AM may assist in overcoming these hurdles by allowing for the on-site creation of critical materials, thereby improving disaster response and recovery operations. Finally, AM has the potential to improve education and boost research in a wide range of sectors [13]. AM can be used as a teaching and research tool in fields such as engineering, medicine, and the arts, helping to improve education and advance research in these areas. For example, AM can be used to produce customised and complex models and prototypes, allowing for the rapid prototyping and testing of new designs and ideas that may possibly be patentable; hence, the development of such ideas requires parts that are not readily available in the market, and hence, in the absence of AM technology, any person will need to rely on factories and industrial plants for the manufacturing of such, which will lead to an increase in the cost of the project and may eventually lead to the stoppage of such projects [14]. This can expedite research and development and enhance education by helping students and researchers more readily envision and grasp complicated concepts and ideas since they will be presented with real-time models that make explanations of their ideas much more possible. Overall, the use of AM has the potential to considerably benefit society by improving access to medical treatment, boosting sustainability, improving disaster response, and improving education and research. AM has the ability to enhance productivity, decrease costs, and enable the creation of personalised and complicated goods, all of which may have a beneficial influence on society and set the example for the development of additional similar technologies while taking prior instances into account [15].

9.4 GROWTH AREA AND MARKET POTENTIAL OF ADDITIVE MANUFACTURING

AM, often known as three-dimensional printing, is a rapidly rising business, with many research analysts anticipating significant market potential in the next decade for this new technology. The global AM market is expected to reach $41.69 billion by 2027, growing at a compound annual growth rate of 24.3% between 2020 and 2027 [16]. Fabrication of end-user parts, components, and utilities has been a prominent development area for AM in recent years. Parts with complex geometry, patterns, manufactured utilities, and internal elements that would be difficult or impossible to duplicate and mass produce using traditional methods may be easily created with AM. As a result, AM has gained popularity in a variety of industries, including aerospace, automotive, medical, and consumer goods. In addition to these applications, there are numerous large-scale opportunities for AM to be used in the production and manufacturing of large-scale structures such as building components, parts, utility items, and infrastructure [17], as well as the production of personalised products such as customised prosthetics and medical implants.

9.5 BIO-BASED RAW MATERIALS IN AM

AM is the technique of constructing a tangible product from CAD. Bio-based raw materials are manufacturing materials produced from biological sources and can be considered non-polluting sources, such as plant-based polymers, rather than fossil fuels. These materials can be used in AM procedures to generate 3D printed products [18]. Polylactic acid (PLA), which is generated from fermented maize starch or sugar cane, is one example of a bio-based raw material used in AM [19]. PLA has qualities similar to petroleum-based polymers and may be used in a range of 3D printing applications, including food packaging, medical equipment, and domestic goods (Figure 9.2). Cellulose, chitosan, and alginate are other bio-based raw materials that have been employed in AM [20].

These materials have the potential to be more sustainable and environmentally friendly than traditional petroleum-based plastics because they can be made from renewable resources and have lower greenhouse gas emissions during production, which is critical given the current market share for such applications, as shown in Figure 9.3. Hemp and mycelium are two materials under consideration for future conversations [21]. These materials have shown promise in recent years, as they have been used to create goods that may be used in everyday activities. Overall, the use of bio-based raw materials in AM is a growing subject with tremendous promise for environmental and economic advantages [19].

9.5.1 Hemp

Hemp is a type of plant in the *Cannabis* genus (*Cannabis* is a genus of flowering plants in the family Cannabaceae, which also includes marijuana (Figure 9.4)). However, hemp and marijuana are different varieties of cannabis that are bred for different purposes. Hemp is grown for its fibres, seeds, and oil, while marijuana is grown for its psychoactive compounds, primarily Tetrahydrocannbinol (THC) (tetrahydrocannabinol). Hemp is typically a tall, thin plant with long, slender leaves. It is an annual plant, meaning it completes its life cycle (from germination to death) within one growing season. Hemp is grown for its fibres; it is a type of plant that is

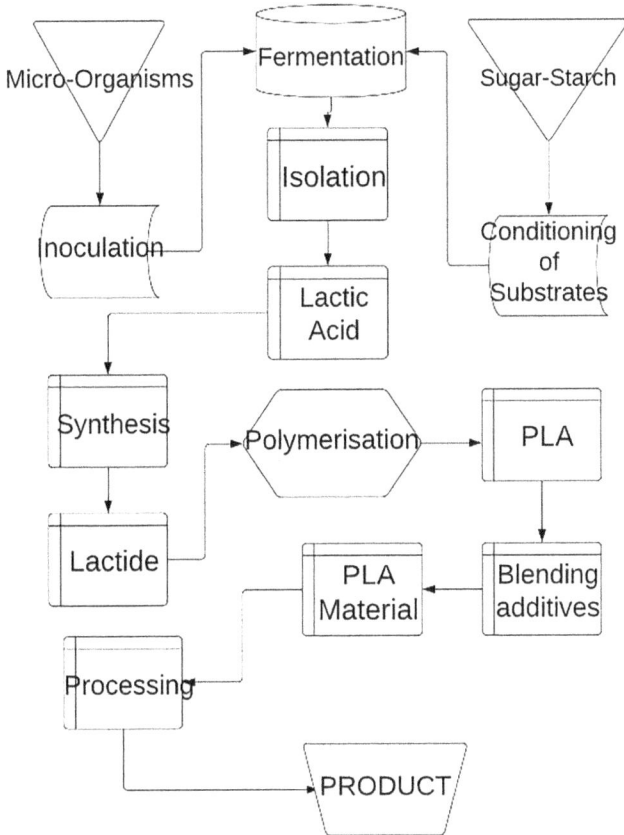

Figure 9.2 PLA production process.

used to produce a variety of products, including textiles, paper, and biofuels [22]. It's also being looked into as a possible raw material for AM, generally known as 3D printing. One possible advantage of employing hemp in AM is that it is a fast-growing, renewable resource that grows with fewer pesticides and fertilisers than other crops. Hemp has a robust and long-lasting fibre structure, making it appropriate for 3D printing applications such as automobile components, construction materials, and industrial items. However, there are significant obstacles to employing hemp in AM. For example, processing hemp into a form appropriate for 3D printing can be challenging, and the mechanical qualities of hemp-based materials may not be as robust as those of typical plastics [23]. In addition, there is still a lack of standardisation and quality control in the production of hemp-based materials for 3D printing, which can make it difficult for manufacturers to consistently produce high-quality products. Overall, hemp has the potential to be a valuable raw material for AM, but further research and development are needed to fully realise this potential [24].

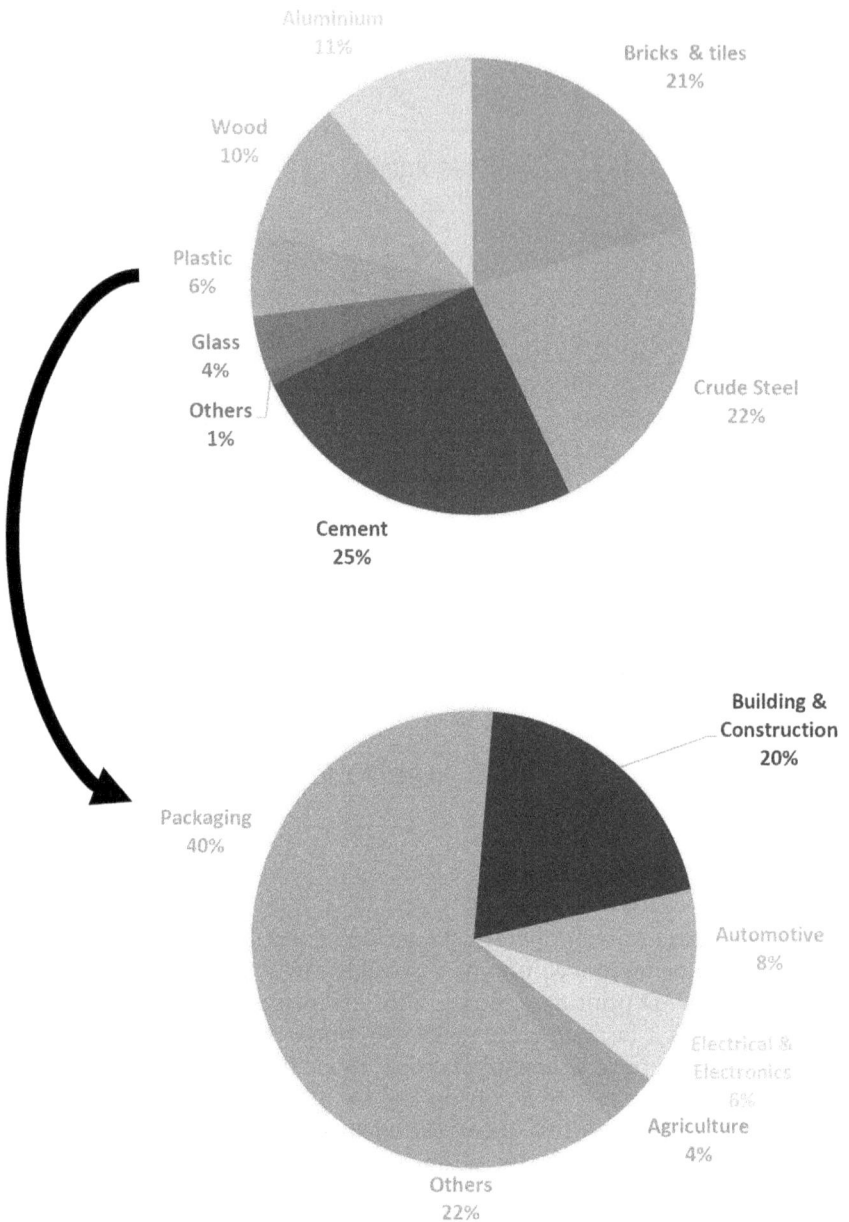

Figure 9.3 Current market share of bulk materials.

9.5.2 Mycelium

Mycelium is a part of the field of microbiology, which is the study of microorganisms such as bacteria, viruses, fungi, and parasites. Within microbiology, mycelium is specifically studied within the branch of mycology,

Figure 9.4 Leaves of hemp.

which is the study of fungi [25]. Mycology is a diverse and important field that encompasses a wide range of subdisciplines, including medical mycology (the study of fungi and their role in human disease), ecological mycology (the study of fungi in their natural environments), and applied mycology (the study of the practical applications of fungi, such as their use in the production of food, medicine, and other products). Overall, the study of mycelium and fungi plays a vital role in understanding the role of microorganisms in the ecosystem and in developing practical applications for these organisms [26]. Mycelium is the vegetative part of a fungus, consisting of a mass of branching, thread-like hyphae that are usually less than 1 millimetre in diameter. Mycelium is responsible for the growth and reproduction of fungi and is the part of the fungus that is typically visible when it is growing in soil or on a substrate. Mycelium has a number of important functions in the ecosystem, including the decomposition of organic matter, nutrient cycling, and forming symbiotic relationships with plants [27]. It is also being explored for its potential use in a variety of applications, including as a sustainable alternative to traditional materials in the production of products such as furniture, building materials, and packaging. In addition, mycelium has been used as a raw material in AM, also known as 3D printing. Mycelium can be grown into a variety of shapes and sizes and has the potential to be used as a biodegradable and renewable alternative to traditional plastics in 3D printing [28]. However, further research is needed to fully understand the potential of mycelium as a raw material in AM and to develop techniques for producing consistent, high-quality products [29].

9.6 HEMP AND MYCELIUM BIO-COMPOSITES

Hemp and mycelium bio-composites refer to the combination of hemp and mycelium, i.e., the composition of biomolecules from hemp and mycelium that could potentially be used in a variety of applications, as both hemp and mycelium have unique properties that make them suitable for different uses in many industrial applications. Hemp is a fast-growing, renewable plant that is known for its strong and durable fibres, which are used to make a variety of products, including textiles, paper, and building materials. It is also grown for its seeds, which are a rich source of protein, fibre, and nutrients and can be used in a variety of food products. Hemp oil, which is extracted from the seeds, is used in a variety of products, including soaps, lotions, and cooking oils. Mycelium is the vegetative part of a fungus, consisting of a mass of branching, thread-like hyphae. It is responsible for the growth and reproduction of fungi and is the part of the fungus that is typically visible when it is growing in soil or on a substrate. Mycelium has a number of important functions in the ecosystem, including the decomposition of organic matter, nutrient cycling, and forming symbiotic relationships with plants. It is also being explored for its potential use in a variety of applications, including as a sustainable alternative to traditional materials in the production of products such as furniture, building materials, and packaging. The potential use of the combination of hemp and mycelium is in the production of bio-composite materials. These materials are made by combining a natural fibre, such as hemp, with a biopolymer, such as mycelium, to create a composite material that has improved properties compared to the individual components. These materials sometimes encompass both positive and negative attributes in their material properties. Bio-composite materials have the potential to be more sustainable and environmentally friendly than traditional materials, as they can be made from renewable resources and have lower greenhouse gas emissions during their production [30]. The required steps to be undertaken to generate a hemp mycelium bio-composite utilising AM techniques can be stated as follows: hemp mycelium powder is mixed with a biopolymer binder, such as chitosan, to form a bio-composite material. After loading the bio-composite material into a 3D printer, the machine is configured to deposit the material layer by layer to build the desired item. The item is allowed to cure after printing, and the biopolymer sets to hold the mycelium particles together. The end product is a bio-composite form of hemp mycelium and a biopolymer binder that may be utilised in a variety of applications, including furniture, packaging, and construction materials. The particular AM process employed will be determined by the available equipment and the needs of the product being manufactured.

9.6.1 Properties of hemp mycelium bio-composites through AM

Hemp and mycelium bio-composites have well-defined and useful mechanical qualities, such as high tensile strength and fatigue resistance under normal circumstances. The mechanical qualities of a hemp and mycelium bio-composite should be determined primarily by the hemp fibre to myce-lium ratio, the processing conditions, and the kind of hemp and mycelium utilised. It should be taken into account. When hemp fibres and mycelium are mixed, they can generate a bio-composite with great tensile strength and fatigue resistance. Hemp and mycelium bio-composites will have excellent tensile strength as well as strong impact resistance and energy absorption. They will also have strong damping qualities, which means that when sub-jected to vibration, they will absorb and disperse energy. Hemp and myce-lium bio-composites often have a low density, making them lightweight materials. As a result, they are suited for use in areas where weight is an issue, such as the automobile sector. Overall, the mechanical qualities of hemp and mycelium bio-composites make them appropriate for use in struc-tural applications as well as industries requiring strength and durability [31].

Hemp and mycelium bio-composites also have good thermal insulation properties, which makes them suitable materials for wide-scale use in the construction industry as they can offer many engineering solutions to certain problems. Usually, the thermal insulation properties of a material are deter-mined by its ability to resist heat transfer through itself. Materials with good thermal insulation properties have a low thermal conductivity, which means they do not conduct heat well. When combined, hemp fibres and mycelium can form a bio-composite with good thermal insulation properties. Using hemp and mycelium bio-composites in construction can help reduce energy consumption in buildings by reducing the amount of heat that is lost through walls, floors, and roofs. This can lead to energy savings and a reduction in greenhouse gas emissions. Overall, the good thermal insulation properties of hemp and mycelium bio-composites make them a sustainable and energy-efficient alternative to traditional insulation materials. Hemp and mycelium bio-composites are quite simple to create and may be moulded into a number of shapes and sizes using a range of methods such as injection moulding, compression moulding, and extrusion. The qualities of the finished bio-com-posite can be affected by processing variables such as temperature, pressure, and humidity. For example, any adjustments made to the thermal parameters during the moulding process, such as raising the temperature and pressure, might result in a stronger and denser bio-composite [32].

9.6.2 Preparation methods of hemp mycelium bio-composites through AM

As there are numerous ways in which AM can be implemented, the types of AM techniques that we will be putting to use to configure hemp mycelium bio-composite are fused deposition modelling (FDM), stereolithography (SLA), selective laser sintering (SLS), binder jetting, and material jetting. FDM technology can be used in AM to create hemp mycelium bio-com-posite. This method creates an item by melting a thermoplastic substance and applying it layer by layer. In this scenario, hemp mycelium powder is combined with a biopolymer binder like chitosan to form a bio-composite material. After that, the bio-composite material is fed into the 3D printer and utilised as the printing medium. The heated nozzle of the 3D printer melts the bio-composite material and deposits it layer by layer according to the design of the desired product. The process is guided by computer software, which divides the design of the item into layers and creates code to direct the printer's motions. After the printing process is finished, the item is allowed to cure for a certain amount of time to allow the biopolymer binder to harden and the mycelium particles to grow together. The end product is a bio-composite consisting of hemp mycelium and a biopolymer binder that has biodegradability, lightweight, and thermal insulating qualities.

The second approach for creating hemp mycelium bio-composite through AM includes the use of SLA technology, which creates an item by curing a liquid resin with a UV laser. In this scenario, hemp mycelium powder is combined with a UV-curable photosensitive resin, such as an acrylate-based resin or epoxy resin, to form a bio-composite material. The SLA 3D printer cures the liquid resin layer by layer, according to the design of the item, using a computer-controlled UV laser. Photo-polymerisation, which occurs when the resin is subjected to UV light, is used by the laser to selectively solidify the resin. The hardened resin shapes the product, while the uncured resin drains away or remains as a support structure. The item is cleaned and cured after printing to allow the resin to harden and the mycelium particles to grow together. The end product is a bio-composite consisting of hemp mycelium and cured resin that has great accuracy, a smooth surface finish, and high strength.

The third AM approach for creating hemp mycelium bio-composite is SLS technology, which is a sort of 3D printing that employs a laser to selectively fuse powder ingredients to make an item. To make a bio-composite material, hemp mycelium powder is combined with a thermoplastic powder, such as PLA or polyamide (PA). The SLS 3D printer employs a computer-controlled laser to selectively fuse the bio-composite powder layer by layer, depending on the design of the product. The laser selectively warms the powder to just below its melting point, causing it to fuse together and form the shape of the item. The unfused powder serves as a support structure and may be utilised in subsequent prints. The item is thereafter cooled and removed from the build chamber after the printing process. After removing the excess powder, the product is post-processed to allow the thermoplastic binder to cure and the mycelium particles to grow together. The end product is a bio-composite consisting of hemp mycelium and a thermoplastic binder that has exceptional strength, durability, and heat resistance.

The fourth AM method for producing hemp mycelium bio-composite is binder jetting, i.e., a 3D printing process that uses a powder bed to deposit a binder material over consecutive layers of bio-composite powder to make a product. The hemp mycelium powder is blended with a binder ingredient, which can be a water-based glue or a bio-based polymer. Binder jetting is a multi-step procedure that begins with the production of a digital model of the thing to be printed. The model is split into thin layers, and a layer of bio-composite powder is deposited onto a construction platform by the

3D printer. Using inkjet printhead technology, the printer then selectively deposits the binder material over the powder layer to make the required shape. The binder material moistens the powder particles, causing them to adhere and form an object. Layer by layer, the process is repeated until the final object is formed, with each layer adhering to the previous one. During the printing process, the unfused powder acts as a support structure and can be reused for subsequent prints. Following the completion of the printing process, the object is post-processed to allow the binder material to set and the mycelium particles to grow together. This involves drying the object in an oven or exposing it to a controlled humidity environment to promote mycelium growth and bonding between the particles. The resulting object is a bio-composite made of hemp mycelium and a binder material that exhibits high strength, durability, and biodegradability.

The fifth AM method for producing hemp mycelium bio-composite is material jetting, i.e., a 3D printing process that deposits microscopic droplets of material onto a build platform, layer by layer, to construct an object. The material used in hemp mycelium bio-composite can be a combination of hemp mycelium particles and a binder substance, such as a water-based glue or a bio-based polymer. The construction of a digital model of the thing to be printed is the first step in the material jetting process. The model is divided into thin layers, and the printer's software directs the inkjet printheads to deposit material droplets on the build platform in accordance with the design of the item. The inkjet printheads deposit material droplets in exact places, layer by layer, constructing the product. Droplets as tiny as a few microns in diameter may be formed, providing excellent precision and accuracy in the finished item. Following the completion of the printing process, the item is post-processed to allow the binder material to solidify and the mycelium particles to grow together. This entails drying the object in an oven or exposing it to a controlled humidity environment in order to stimulate mycelium development and particle attachment. The end product is a bio-composite composed of hemp mycelium and a binder ingredient with great strength, durability, and biodegradability.

9.7 MYCELIUM LEATHER: ALTERNATIVE TO ANIMAL LEATHER

Mycelium leather is a biodegradable substance created from the fibrous network of mycelium, the fungus vegetative portion. Mycelium creates a thick network of interlocking fibres on agricultural waste such as sawdust or maize stalks [33]. The fibres are then compacted, dried, and tanned to produce a leather-like material that is both tough and flexible (Figure 9.5). Mycelium leather is a non-toxic, sustainable alternative to regular leather that may be made in a variety of colours and textures. It might be used in the fashion, interior design, and automobile industries, where sustainable materials are becoming more significant [34].

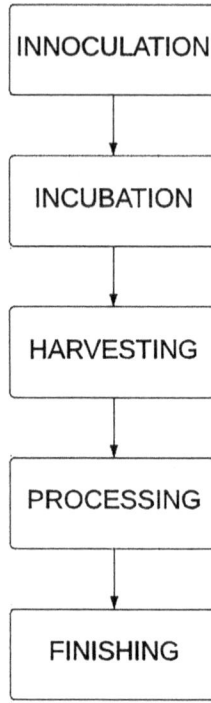

Figure 9.5 Mycelium leather creation process.

9.7.1 Process for creation of mycelium leather

The process for creating mycelium leather involves the following steps:

Inoculation: Mycelium is grown from spores that are mixed with a growth medium, such as straw or sawdust. The spores germinate, and the mycelium begins to grow and colonise the growth medium.

Incubation: The mycelium is allowed to grow and colonise the growth medium in a controlled environment, such as a laboratory or greenhouse. The temperature, humidity, and other conditions are carefully regulated to optimise the growth of the mycelium [35].

Harvesting: Once the mycelium has fully colonised the growth medium, it is harvested and dried. The mycelium can be harvested in a variety of forms, including sheets or blocks.

Processing: The harvested mycelium is then processed to form a leather-like material. This can involve a variety of steps, such as cutting, pressing, and tanning. The specific processing steps will depend on the desired properties and appearance of the final mycelium leather.

Finishing: The processed mycelium leather is then finished to give it the desired appearance and properties. This can involve a variety of steps, such as dying, embossing, and coating.

Overall, the process for creating mycelium leather involves growing and harvesting the mycelium, and then processing and finishing it to form a leather-like material [36].

9.8 HEMP AS A THERMAL RESISTANT MATERIAL

Hemp is a versatile plant that has a number of applications in the textile industry. Hemp fibres are strong, durable, and resistant to UV radiation, making them suitable for use in a variety of textile products. They are also naturally hypoallergenic. (Hypoallergenic refers to materials or products that are less likely to cause an allergic reaction in people who are sensitive to certain substances.) Hypoallergenic products are designed to minimise the presence of allergens (or to reduce the risk of an allergic reaction) and have moisture-wicking properties, making them comfortable to wear and sleep on [37]. Hemp fabrics are also sustainable and environmentally friendly. Hemp is a fast-growing plant that requires little water and pesticides, and the production of hemp fabrics generates fewer greenhouse gas emissions compared to the production of synthetic fabrics. In addition to its practical and environmental benefits, hemp also has a natural, earthy look that is appreciated by many consumers [38]. Hemp fabrics are available in a range of colours, from natural shades of beige and brown to dyed shades of blue, green, and red. Keeping all this in context, we can consider hemp a suitable material for use in a variety of textile products, including bed sheets, clothing, aprons, curtains, and other household items. Its strength, durability, and sustainability make it a popular choice for both practical and environmental reasons [39].

9.8.1 Process of creation of household textiles using hemp

The process for creating bed sheets, clothing, aprons, curtains, and other household items using hemp involves the following steps (Figure 9.6):

Cultivation: Hemp plants are grown and harvested for their fibres, which are used to make textiles. Hemp is a fast-growing plant that requires little water and pesticides and can be grown in a variety of climates [40].

Retting: After the hemp plants are harvested, the fibres are separated from the plant's woody core through a process called retting. Retting can be done through various methods, such as water retting or dew retting.

Breaking and scutching: The retted hemp fibres are then broken and scutched, which involves removing the remaining woody core and impurities from the fibres [40].

Carding and spinning: The cleaned fibres are then carded, which involves aligning the fibres in the same direction, and spun into yarn. The yarn can be spun into various thicknesses, depending on the desired end use.

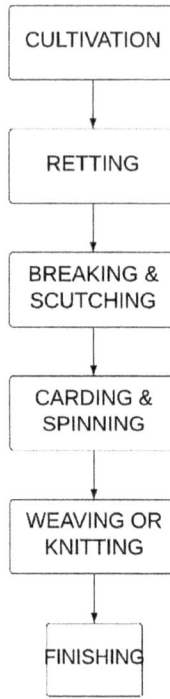

Figure 9.6 Handloom, textile creation using hemp.

Weaving or knitting: The yarn is then woven or knitted into fabric using a variety of techniques, such as plain weave, twill weave, or jersey knit [40]. The fabric can be woven or knitted into a variety of weights, textures, and finishes, depending on the desired end use.

Finishing: The fabric is then finished to give it the desired appearance and properties. This can involve a variety of steps, such as dyeing, printing, or softening [41].

The material jetting process is to be used to manufacture mycelium leather through AM. In this procedure, an inkjet printhead is loaded with a combination of mycelium and a binder substance, such as a bio-based polymer. The printer then ejects droplets of the mixture, layer by layer, onto a build platform to produce a 3D object. After that, the item is post-processed in a regulated humidor environment to allow mycelium to grow and form a dense network of fibres. After the mycelium has grown, the item is crushed, dried, and tanned to create pliable and durable mycelium leather.

Using hemp as a production material for AM to manufacture bed sheets, garments, aprons, curtains, and other home products. The procedure begins with the object being designed using CAD software. The design is then translated into a format that the AM machine can understand [42]. The AM machine then deposits layers of a hemp-based substance, such as hemp plastic or hemp-infused filament, onto a build platform to create the 3D item. The item might be printed in one piece or in portions that are later put together. When the printing is finished, the item is removed from the machine, and any surplus material is removed [43]. Depending on the type of item being manufactured, extra post-processing operations such as sanding, polishing, or painting may be required. To complete clothing items, extra processes such as cutting, stitching, and attaching fasteners may be necessary. Using AM to make hemp home goods has various advantages. The method is extremely adjustable, allowing for the creation of one-of-a-kind patterns and forms. It also provides a more sustainable and environmentally friendly alternative to traditional production processes, requiring less energy and producing less waste. However, it is crucial to highlight that the technology for making clothes and other soft items is still in its early phases of development and that more study is needed to improve the process for these applications.

REFERENCES

[1] Ponni, R. T., Swamivelmanickam, M., & Siva Krishnan, M. R. (2020). 3D printing in pharmaceutical technology - a review. *International Journal of Pharmaceutical Investigation*, 10(1): 8–12.

[2] Goyanes, A., Madla, C. M., Umerji, A., Pineiro, G. D., Giraldez Montero, J.M., Lamas Diaz, M.J., Barcia, M. G., Taherali, F., Pintos, P. S., Couce, M. L., Gaisford, S., Basit, A. W. (2019). Automated therapy preparation of isoleucine formulations using 3D printing for the treatment of MSUD: first single-centre, prospective, crossover study in patients. *International Journal of Pharmaceutic*, 567: 118497.

[3] Leu, M. C., & Lu, Y. C. (2017). Additive manufacturing in aerospace: a review. *Aerospace Science and Technology*, 71: 359–369.

[4] Smith, S. L., Mitchell, R. T., & Campbell, S. A. (2017). Additive manufacturing in aerospace: a review of materials and applications. *Additive Manufacturing*, 31: 1–12.

[5] Alvis, J. M., & Coker, M. L. (2018). Additive manufacturing in aerospace: a review of opportunities and challenges. *Aerospace Science and Technology*, 31: 446–454.

[6] Gebler, M., Uiterkamp, A. J. M. S., & Visser, M. R. (2014). A global sustainability perspective on 3D printing technologies. *Energy Policy*, 74: 158–167.

[7] Hager, I., Golonka, A., & Putanowicz, R. M. R. (2016). 3D printing of buildings and building components as the future of sustainable construction. *Procedia Engineering*, 151: 292–299.

[8] El-Sayegh, S., Romdhane, L. & Manjikian, M. R. (2020). A critical review of 3D printing in construction: benefits, challenges, and risks. *Archives of Civil and Mechancial Engineering*, 20: 34.

[9] The authority on 3D printing & additive manufacturing. (2022, July 12). *SME Business Ideas, News, Tips & Latest Updates - TradeIndia*. https://www.tradeindia.com/blog/authority-on-3d-printing-and-additive-manufacturing/.

[10] Volvo trucks improves quality with 3D printing technology at its New River Valley plant. (2019, September 4). *Volvogroup.com*. https://www.volvogroup. com/en/news-and-media/news/2019/sep/printing-technology.html.

[11] He, C., Zhang, M., & Fang, Z. (2020). 3D printing of food: pretreatment and post-treatment of materials. *Critical Reviews in Food Science and Nutrition*, 60(14): 2379–2392.

[12] Suderman, A. M., & Jablonsky, D. L. (2018). 3D printing and society: an exploration of promising applications and implications. *Technological Forecasting and Social Change*, 10: 1016.

[13] Thiesse, F., Wirth, M., Kemper, H. G. et al. (2015). Economic Implications of Additive Manufacturing and the Contribution of MIS. *Business & Information Systems Engineering* 57: 139–148. https://doi.org/10.1007/s12599-015-0374-4

[14] Huang, S.H., Liu, P., Mokasdar, A. et al. (2013). Additive manufacturing and its societal impact: a literature review. *The International Journal of Advanced Manufacturing Technology*, 67: 1191–1203. https://doi.org/10.1007/ s00170-012-4558-5.

[15] KarelKellens, R. M., & DimosParaskevas, W. D. (2017). Environmental impact of additive manufacturing processes: does AM contribute to a more sustainable way of part manufacturing?. *Procedia CIRP*, 61: 582–587.

[16] Tang, S., Yang, L., Fan, Z., Jiang, W., & Liu, X. (2021). A review of additive manufacturing technology and its application to foundry in China. *China Foundry*, 18: 249–264. Doi: 10.1007/s41230-021-1003-0.

[17] Ashutosh Tiwari, D. S. M., & Davim, J. P. (2017). Additive manufacturing: opportunities for growth in the automotive industry. *Materials and Manufacturing Processes*, 10: 1080.

[18] Siwal, S.S., Mishra, K., Saini, A.K. et al. (2022). Additive manufacturing of bio-based hydrogel composites: recent advances. *Journal of Polymers and the Environment*, 30: 4501–4516. https://doi.org/10.1007/s10924-022-02516-z

[19] Van Wijk, A. J. M., & van Wijk, I. (2015). *3D Printing with Biomaterials: Towards a Sustainable and Circular Economy*. Amsterdam, Netherlands: IOS press.

[20] Kim, Y. G., Kim, J. H., & Yeon-Sik Jung, M. R. (2020). Additive manufacturing of bio-based polymer composites: a review. *Composites Part B: Engineering*, 2019: 107607.

[21] Bedian, L., Villalba-Rodríguez, A. M., Hernández-Vargas, G., Parra-Saldivar, R., & Iqbal, H. M. N. (2017). Bio-based materials with novel characteristics for tissue engineering applications – A review. *International Journal of Biological Macromolecules*, 98: 837–846. https://doi.org/10.1016/j. ijbiomac.2017.02.048

[22] Small, E., & Marcus, M. R. (2002). Hemp: a new crop with new uses for North America. *Trends in New Crops and New Uses*, 24(5): 284–326.

[23] Kozlowski, R. M., & Mackiewicz-Talarczyk, M. (2020). *Handbook of Natural Fibres: Volume 1: Types, Properties and Factors Affecting Breeding and Cultivation*. Woodhead Publishing.

[24] Shahzad, A. (2012). Hemp fiber and its composites-a review. *Journal of composite materials*, 46(8): 973–986.

[25] Bibbins-Martínez, M., Juárez-Hernández, J., López-Domínguez, J. Y., Nava-Galicia, S. B., Martínez-Tozcano, L. J., Juárez-Atonaĺ, R., Cortés-Espinosa, D., & Díaz-Godinez, G. (2023), Potential application of fungal biosorption and/or bioaccumulation for the bioremediation of wastewater contamination: A review. *Journal of Environmental Biology*, 44(2): 135-145.

[26] Li, K., Jia, J., Wu, N., & Xu, Q. (2022). Recent advances in the construction of biocomposites based on fungal mycelia. *Frontiers in Bioengineering and Biotechnology*, 10. https://doi.org/10.3389/fbioe.2022.1067869

[27] Sommano, S. R., Suksathan, R., Sombat, T., Seehanam, P., Sirilun, S., Ruksiriwanich, W., Wangtueai, S., & Leksawasdi, M. R . (2022). Novel perspective of medicinal mushroom cultivations: a review case for 'magic' mushrooms. *Agronomy*, 12(12): 3185.

[28] Hoa, H. T., & Wang, M. R. (2015). The effects of temperature and nutritional conditions on mycelium growth of two oyster mushrooms (Pleurotus ostreatus and Pleurotus cystidiosus). *Mycobiology*, 43(1): 14–23.

[29] Islam, M. R., Tudryn, G., Bucinell, R., Schadler, L., & Picu, R. C. (2017). Morphology and mechanics of fungal mycelium. *Scientific Reports*, 7(1): 13070.

[30] Zimele, Z., Irbe, I., Grinins, J., Bikovens, O., Verovkins, A., & Bajare, D. (2020). Novel mycelium-based biocomposites (Mbb) as building materials. *Journal of Renewable Materials*, 8(9): 1067–1076.

[31] Ziegler, A. R., Bajwa, S. G., Holt, G. A., McIntyre, G., & Bajwa, D. S. (2016). Evaluation of physico-mechanical properties of mycelium reinforced green biocomposites made from cellulosic fibers. *Applied Engineering in Agriculture*, 32(6): 931–938.

[32] Butu, A., Rodino, S., Miu, B., & Butu, M. (2020). Mycelium-based materials for the ecodesign of bioeconomy. *Digest Journal of Nanomaterials and Biostructures*, 15(4): 1129–1140.

[33] Javadian, A., Le Ferrand, H., Hebel, D. E., & Saeidi, N. (2020). Application of mycelium-bound composite materials in construction industry: a short review. *SOJ Materials Science and Engineering*, 7: 1–9.

[34] Bitting, S., Derme, T., Lee, J., Van Mele, T., Dillenburger, B., & Block, P. (2022). Challenges and opportunities in scaling up architectural applications of mycelium-based materials with digital fabrication. *Biomimetics*, 7(2): 44.

[35] Bustillos, J., Loganathan, A., Agrawal, R., Gonzalez, B. A., Perez, M. G., Ramaswamy, S., & Agarwal, A. (2020). Uncovering the mechanical, thermal, and chemical characteristics of biodegradable mushroom leather with intrinsic antifungal and antibacterial properties. *ACS Applied Bio Materials*, 3(5): 3145–3156.

[36] Williams, E., Cenian, K., Golsteijn, L., Morris, B., & Scullin, M. L. (2022). Life cycle assessment of MycoWorks' Reishi(tm): the first low-carbon and biodegradable alternative leather. *Environmental Sciences Europe*, 34(1): 120.

[37] Sathishkumar, T. P., Naveen, J. A., & Satheeshkumar, S. (2014). Hybrid fiber reinforced polymer composites-a review. *Journal of Reinforced Plastics and Composites*, 33(5): 454–471.

[38] Saheb, D. N., & Jog, J. P. (1999). Natural fiber polymer composites: a review. *Advances in Polymer Technology: Journal of the Polymer Processing Institute*, 18(4): 351–363.

[39] Kaur, G., & Kander, R. (2023). The sustainability of industrial Hemp: A literature review of its economic, environmental, and social sustainability. *Sustainability*, 15, 6457. Doi: 10.3390/su15086457

[40] Manaia, J. P., Manaia, A. T., & Rodriges, L. (2019). Industrial hemp fibers: an overview. *Fibers*, 7(12): 106.

[41] Dhondt, F., & Muthu, S. S. (2021). *Hemp and Sustainability*. Berlin/Heidelberg, Germany: Springer.

[42] Ahmed, A. F., Islam, M. Z., Mahmud, M. S., Sarker, M. E., & Islam, M. R. (2022). Hemp as a potential raw material toward a sustainable world: a review. *Heliyon*, 8: e08753.

[43] Bozsaky, D. (2019). Nature-based thermal insulation materials from renewable resources-a state-of-the-art review. *Slovak Journal of Civil Engineering*, 27(1): 52–59.

Chapter 10

Characterisation of titanium feedstock powder prepared by recycling of machining chips using ball milling

Prameet Vats, Arun Kumar Bambam,
Kishor Kumar Gajrani, and Avinash Kumar
Indian Institute of Information Technology,
Design and Manufacturing

10.1 INTRODUCTION

Titanium-based metals are extensively used in industries such as aircraft, automobiles, and medicine because of their good mechanical properties, resistance to rust, and interaction with living organisms. However, the expensive cost of producing titanium restricts its wide-scale usage. As a result, recycling titanium chips has become a vital option for reducing manufacturing costs and minimising environmental effects [1]. Machining chips (MCs) produced during the cutting of titanium alloy-based components can be a great source of titanium waste for recycling. Traditionally, scrap (size 1,000 µm) is recycled using a variety of processes, including crushing, centrifugation (oil recovery), melting in furnaces, thermal degreasing, and casting into ingots. There are numerous constraints on the melting step of recycling. These include the use of water-cooled copper crucibles, vacuum and inert environment requirements, as well as energy-intensive vacuum arc melting furnaces. Furthermore, in order to ensure homogeneity in the recovered material, this method can only melt a very small quantity of chips. Metal recycling through these processes uses more than 6,000 L of water per metric tonne of output [2]. Whenever the chips are melted, they release metallic smoke because of the oxidation of the impurities. The smoke has a high global warming potential and is hazardous for the environment. All of these difficulties add up to a permanent metal loss (15%–25%), rendering the whole process unsustainable [3]. To overcome these issues, conventional methods of recycling such as spark plasma [4], cold pressing [5], and hot extrusion [6] have been investigated. However, these solid-state manufacturing techniques have been observed to have difficulties with poorer mechanical properties and poor surface topography of the produced components [7]. Therefore, to get a finer shape and size with good mechanical properties, ball milling was introduced.

DOI: 10.1201/9781003406488-10

The use of ball milling to recycle titanium MCs has attracted substantial interest in recent years owing to its cheap cost, simplicity, and efficacy. The ball milling method employs high-energy mechanical forces to break down titanium chips into smaller particles that may be used as feedstock powder in the manufacturing of titanium alloys [8,9]. The characterisation of titanium feedstock powder created by recycling MCs via ball milling is critical for ensuring the end product's quality and performance. The particle size distribution, shape, chemical content, and crystalline structure of the feedstock powder are all examined throughout the characterisation process [10]. These qualities govern the behaviour of the feedstock powder throughout subsequent processing steps such as powder compaction and sintering, which impact the mechanical and chemical properties of the end product. Figure 10.1 illustrates the flow of recycled MCs into powder for utilisation.

This chapter aims to provide an overview of the characterisation of titanium feedstock powder manufactured by recycling MCs using ball milling. The study comprises optimising ball milling parameters to produce a feedstock powder with suitable characteristics. The feedstock powder characterisation data are evaluated using several methods, such as X-ray diffraction (XRD), scanning electron microscopy (SEM), transmission electron microscopy (TEM), and energy dispersive spectroscopy (EDX).

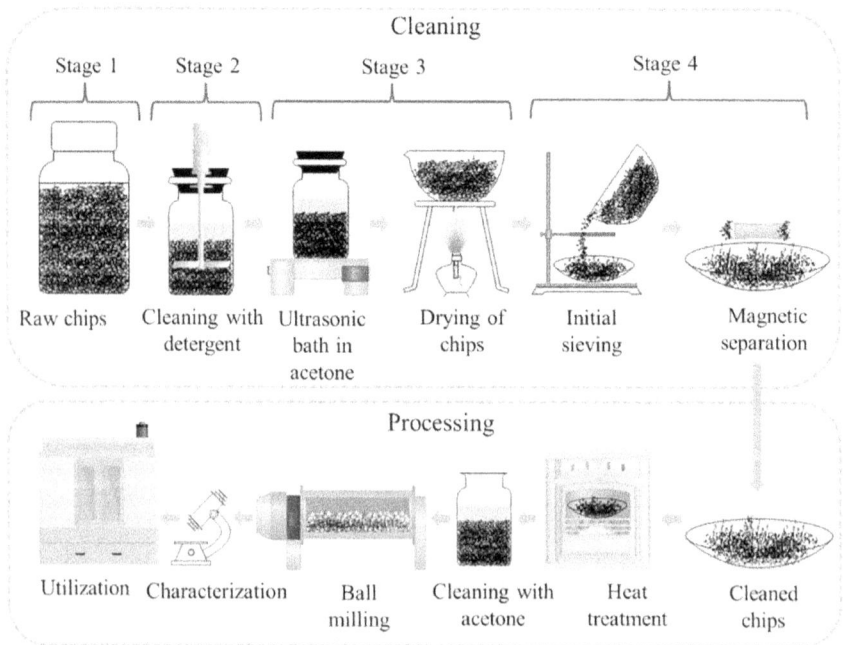

Figure 10.1 Process of recycling machining chips [9].

The acquired findings give important insights into the characteristics of the feedstock powder and may be used to guide the selection of optimum processing settings for the manufacturing of high-quality titanium alloys.

10.2 SUSTAINABLE ISSUES AND CHALLENGES

To achieve sustainability, industries must strive towards processes that use less energy, produce fewer emissions, and create less waste. To ensure the overall sustainability of the process, waste should be controlled using a waste management hierarchy at the intermediate and, most significantly, final stages of the product life cycle [11]. Metal cutting using conventional subtractive machining processes is a necessary operation in the technical design and manufacturing sectors. To fulfil contemporary technological demand, traditional subtractive machining processes are frequently employed to make engineering components. The biggest disadvantage for these sectors is their massive waste in the form of lubricant, energy, and material waste as chips [12]. As a result, the titanium sector has encountered sustainability challenges in terms of raw material pricing and availability, energy consumption, environmental impact, and waste management. The industries are concerned about the high cost of raw materials and energy use. Furthermore, the extraction of titanium from its ores generates significant waste, necessitating adequate waste management and disposal systems. During the extraction of titanium ores, a significant amount of carbon dioxide is released, which contributes to global warming. Mining for titanium ores also pollutes the environment due to the release of heavy metals and other impurities [13]. By considering the scenario and issue rerated, titanium recycling of MCs is one of the approaches to decreasing the environmental impact of titanium in industries. Recycling titanium chips is a critical solution to sustainability problems. It lowers raw material costs, energy usage, and trash generation. Furthermore, recycling of MCs minimises the titanium industry's environmental impact by reducing carbon dioxide emissions and the requirement for new titanium ores.

10.3 CURRENT SCENARIO OF MACHINING CHIPS RECYCLING INDUSTRY

Metal chip recycling is the collection and processing of waste created during machining. The MC recycling industry is now in good shape, with an increasing number of firms providing recycling services for different metals, including titanium. Growing demand for recycled feedstock powders is being driven by increasing raw material prices, restricted resource availability, expenses, and environmental issues connected with the extraction and processing of new materials. Recycling metal chips is crucial because it lowers the cost of raw resources, trash production, and energy use [14,15]. Whereas metal chip recycling comes with a number of problems, including collection and transportation, separation from other waste materials, and processing into usable products [16]. The recycling of titanium chips is a new sector that has gained traction as a result of environmental concerns in the industry. The titanium chip recycling sector is still in its early stages, and more research is needed to increase the recycling process's efficiency and efficacy. MCs have the ability to be used as raw materials in the creation of innovative technical products. It is estimated that MCs account for a significant portion (13.7% aluminium and 14.6% steel) of total industrial waste created worldwide [17]. However, when we examine the proportion of titanium (a difficult-to-machine material), it is predicted that 55% of the entire input material is transformed into MCs (national statistics from the United States) [18] (see Figure 10.2).

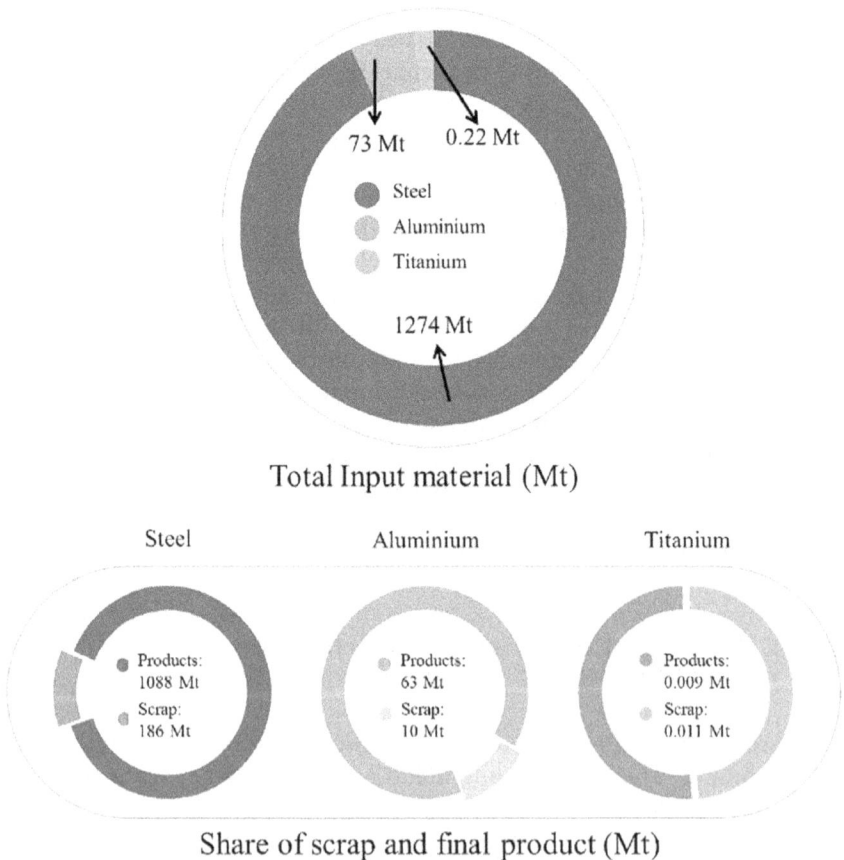

Total Input material (Mt)

Share of scrap and final product (Mt)

Figure 10.2   Current statistics of machined chips for the recycling industry [17].

The desire for sustainable solutions is propelling the expansion of the MC recycling business, and governments throughout the globe are enacting rules and regulations to encourage metal scrap recycling. The European Union's Circular Economy Action Plan intends to encourage sustainable production and consumption by lowering waste output and promoting recycling [19]. Furthermore, the circular economy movement is promoting the use of closed-loop supply chains, in which scrap materials are recovered and repurposed to make new components. This trend is projected to propel the MC recycling sector even further.

10.4 RECYCLING OF CHIPS

Titanium chip recycling is the collection, separation, and conversion of MCs into feedstock powder for powder metallurgy. The powder feedstock is utilised to create components with the required form, size, and characteristics [20]. The benefits of recycling titanium chips include lower waste generation, lower raw material costs, and a lower environmental impact. The drawbacks of recycling titanium chips are the high processing costs and the difficulties in creating feedstock powder with the appropriate characteristics [9].

There are numerous ways to recycle titanium chips, such as vacuum sintering, plasma sintering, and ball milling. Two approaches for producing bulk components directly from MCs are vacuum sintering and plasma sintering. These procedures, however, need high processing temperatures, which might result in chemical reactions and contamination of the feedstock powder [21]. Furthermore, the presence of contaminants in the sintered components may result in worse mechanical qualities. In recent studies, it was observed that MCs were recycled and used as a feedstock material in the additive manufacturing process. This process was carried out by means of solidification and an energy-intensive melting process known as additive friction stir deposition (AFS-D) [22]. The results demonstrated the feasibility of recycling MCs by directly feeding them into the AFS-D process and producing solid structural deposition that can be used for point-of-need production, especially in harsh environments where multi-step processing of MCs is not possible due to the spherical shape of the powder. Thus, the powder generated by gas atomisation is the most common feedstock among the aforementioned conventional powder production techniques (CPPTs), owing to the controlled distribution of particle size and spherical shape [23].

While it is crucial to note that the powder made from CPPTs consumes a large amount of energy, the end product costs several times more. As a consequence, CPPTs are only cost-effective for mass manufacture. Furthermore, owing to the economics of the processes, CPPTs are not practicable for powder manufacturing from a range of alloys that are not

Equal channel angular pressing (ECAP)	Melting
1. Limited processing direction 2. Mass prodroction is impossible 3. High energy input 4. Sophisticated equipment	1. Generation of high temperature 2. Metal loss due to oxidation 3. Inclusion of impurities 4. Metal recovery is only 50-60%
Issues with direct processes for reycling of machining chips	
Extrusion	Forging
1. Simple shape are possible 2. Wide facility area is required 3. Limited processing direction 4. Inferior properties of parts	1. Partial cracks due to shear stress 2. Cost of mold is expensive 3. High energy input 4. Frequent wear out of molds

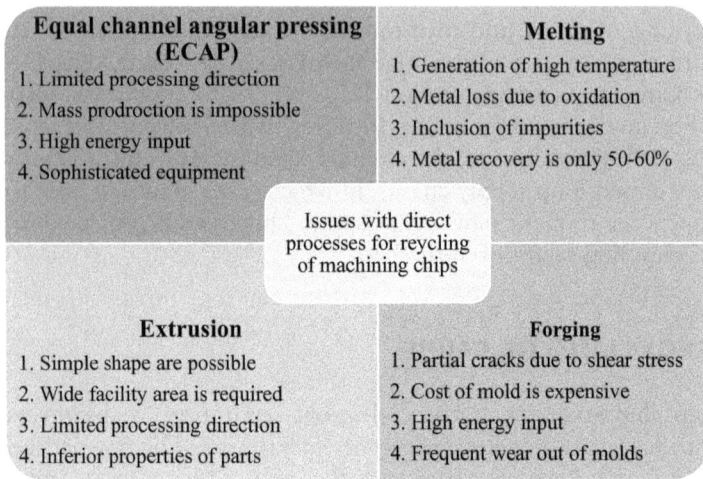

Figure 10.3 Limitation of the direct conventional process for recycling machining chips.

typically utilised [24]. Figure 10.3 lists the issues with direct conventional processes for recycling MCs. To address these problems, ball milling has lately been investigated as a feasible method for MC powder manufacture. Ball milling is a popular process for converting MCs into powder feedstock. It entails mechanical deformation and fracture of particles, leading to the development of new surfaces and particle size reduction.

10.5 BALL MILLING

MCs are often seen as trash, although they may be recycled and used as a feedstock in the manufacture of titanium powder. The chips are mechanically ground in a ball mill during this operation [8]. Ball milling is a mechanical process that uses a spinning container packed with balls made of different materials such as steel, ceramic, or glass. The material to be ground is placed into the container and exposed to the mechanical action of the balls, which results in particle size reduction. The mechanical energy created by collisions between the balls and the particles of the MCs produces deformation, fragmentation, and plastic deformation of the chips during ball milling. Heat is also generated throughout the operation, which might impact the characteristics of the resultant powder [25]. To obtain the appropriate particle size, morphology, and characteristics, milling parameters such as the ball-to-powder ratio, milling duration, rotation speed, and ball type and size must be optimised [26]. Researchers divide ball mills into two types based on their way of operation: direct and indirect. In the case of direct ball milling, rollers or mechanical shafts work directly on the particles and transmit kinetic energy. In the case of indirect ball milling, the kinetic energy is transferred first to the device's body, then to the grinding medium. There have been various methods of ball milling developed throughout the years, such as tumbler ball milling, planetary ball milling, attrition milling, and vibratory ball milling (see Figure 10.4) [25,27].

10.5.1 Tumbler ball milling

Tumbler ball milling is a method of mixing materials that uses a spinning drum packed with metal or ceramic balls to break down and mix them together effectively. It is widely used in the manufacture of ceramics, electronics, and medicines. The materials and balls are fed into the drum and spun at a regulated speed. The process duration varies according to the materials and the intended result [28].

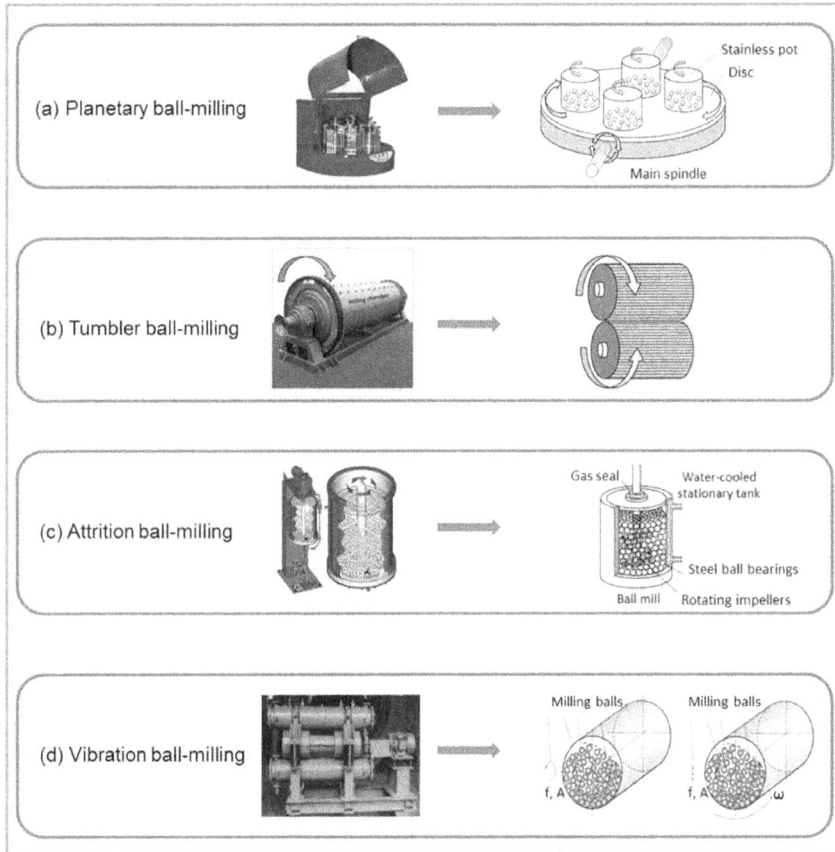

Figure 10.4 Clasification of the ball milling process on the basis of the mill [25].

10.5.2 Planetary ball milling

Planetary ball milling is a form of ball milling in which a planetary ball mill is used to grind and mix materials in controlled conditions. The milling jar in planetary ball milling is positioned on a planetary disc that revolves around its own axis while simultaneously spinning around a central axis. The milling jar balls may bounce back and forth with a high frequency due to the centrifugal force produced by the dual spin, which causes high-energy collisions between the balls and the powder. This method is often used to create nanomaterials and composites [26].

10.5.3 Attrition milling

Attrition milling is a kind of ball milling that grinds and disperses powder using a series of rotating agitator arms. The powder, together with the balls and revolving agitator arms, is put into a grinding chamber in attrition milling. The agitator arms shear the powder, causing it to be crushed and disseminated by the balls. Attrition milling is often used to make fine powders and dispersions [29].

10.5.4 Vibratory ball milling

Vibratory ball milling is a sort of ball milling in which the milling jar and its contents are vibrated using a vibrating mill, which is a specialised equipment intended to vibrate the milling jar and its contents. The powder is loaded into the milling jar together with the balls in vibratory ball milling, and the jar is then placed on a vibrating platform. The vibrations force the balls in the jar to vibrate with high frequency, resulting in high-energy collisions between the balls and the powder. Nanomaterials and composite materials are often produced using vibratory ball milling [30].

10.6 CHARACTERISATION OF BALL-MILLED POWDER

10.6.1 Particle size

The particle size distribution of the ball-milled powder is a key property that influences the sinter ability and feedstock powder qualities. Laser diffraction, dynamic light scattering, SEM, and sedimentation analysis may all be used to estimate the particle size distribution. The particle size distribution might provide information about the powder size and quality [31]. The powder size representation is shown in Figure 10.5.

10.6.2 Phase identification and hardness

To determine the phases present in the ball-milled powder, XRD and SEM are often used. Micro-indentation and Vickers hardness tests may be used to assess the hardness of the ball-milled powder. The hardness of titanium increases linearly with milling time. This is attributed to the reduction in

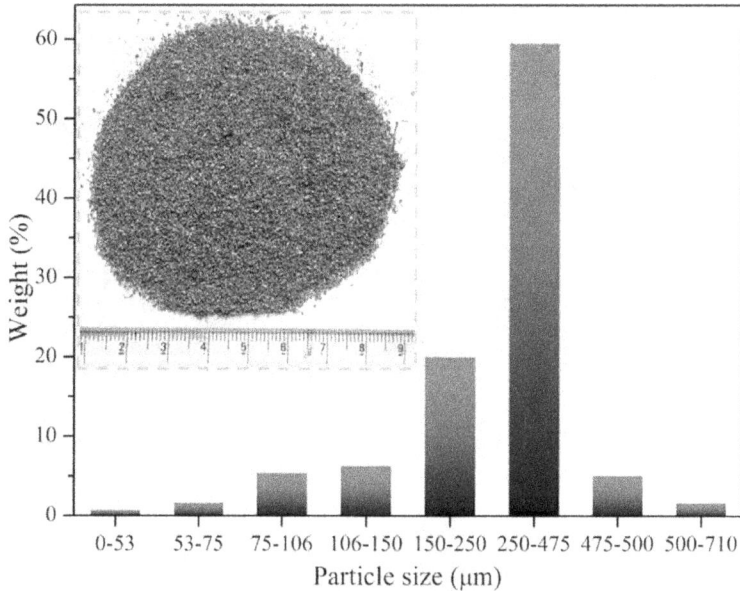

Figure 10.5 Particle size of ball-milled powder [9].

grain size and high dislocation density caused by work hardening due to constant hammering action by ball mills [32].

10.6.3 Powder flowability and spreadability

The flowability and spread ability of ball-milled powder are critical properties that influence powder handling and compaction during powder metallurgy. Techniques such as angle of repose, tapping density, and Carr index may be used to measure flowability and spreadability [33]. Powder rheometry and angle of repose tests may be used to determine the flowability and spreadability of the powder. These qualities are critical for applications such as additive manufacturing, where the powder must flow smoothly to guarantee high-quality components are produced [34].

10.6.4 Morphology and aspect ratio

The ratio and morphology of the ball-milled powder are important criteria that determine sintering behaviour and feedstock powder quality. To determine the morphology and aspect ratio, techniques such as SEM, EDX (see Figure 10.6), and TEM may be utilised. These qualities may provide information on the shape, composition, and structure of the particles, which may influence the final properties [35].

10.6.5 Single tracks

Single tracks produced through laser cladding and powder bed fusion are used to examine the viability of powder for additive manufacturing applications.

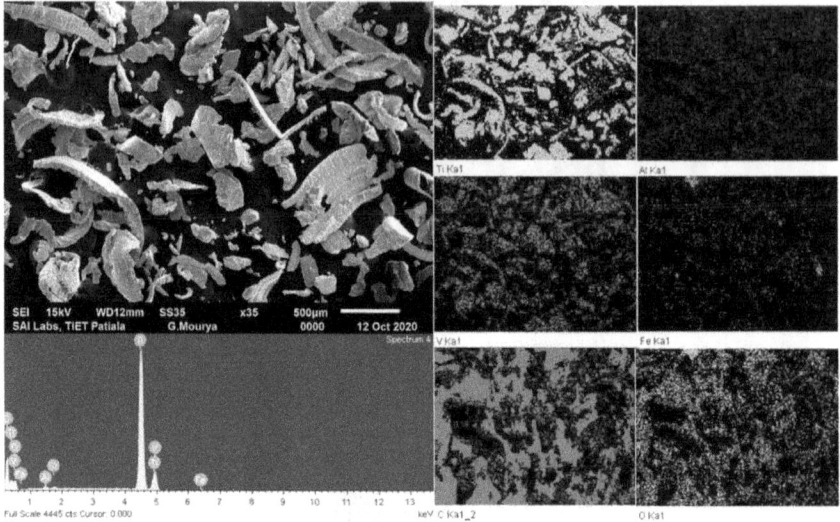

Figure 10.6 Morphology and composition representation via SEM and EDX, respectively [9].

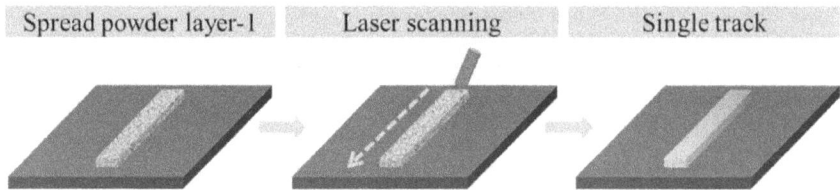

Figure 10.7 Single track representation of prepared powder by using direct metal laser sintering [9].

To analyse the microstructure, fracture development, and mechanical properties of single tracks, techniques including optical microscopy, SEM, and micro-indentation may be utilised [36]. Figure 10.7 shows the steps of single-tracking the prepared powder by using direct metal laser sintering.

10.7 APPLICATIONS

The titanium feedstock powder produced by ball milling recycled MCs has a wide range of possible uses (see Figure 10.8). Some of the applications are discussed below.

10.7.1 Powder metallurgy

The powder may be used to make titanium alloys, which are extensively used in the aerospace and biomedical sectors. Ball milling may increase the flowability, density, and packing qualities of the powder, making it simpler

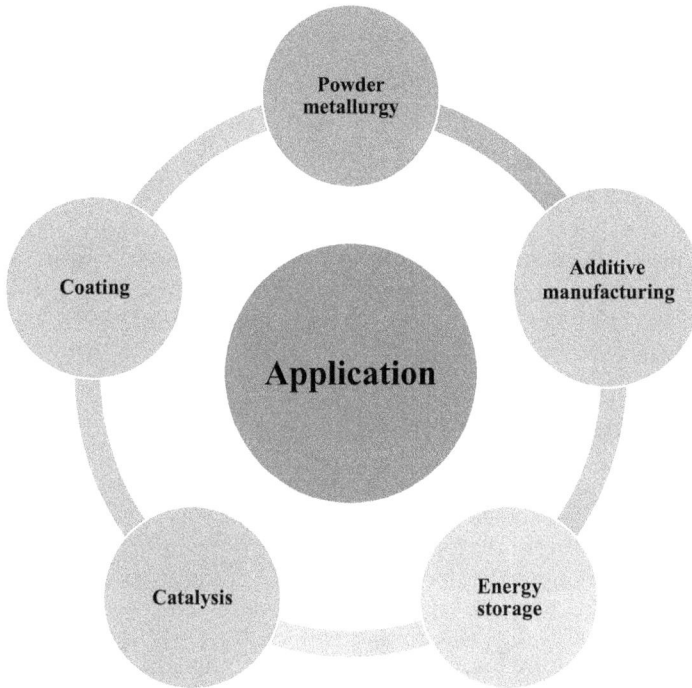

Figure 10.8 Application of titanium feedstock powder produced by ball milling recycled machining chips.

to handle throughout the powder metallurgy process [37]. Furthermore, the customised qualities of the powder might result in increased mechanical and corrosion resistance in the finished product.

10.7.2 Additive manufacturing

Powder may be used as a feedstock in additive manufacturing processes such as powder bed fusion and binder jetting to create complicated shapes. Titanium is an excellent material for additive manufacturing applications owing to its high melting point and resistance to corrosion. The powder produced through the ball milling operation is small in size and exhibits good mechanical properties, which results in a good surface finish, high dimension accuracy, high hardness, and strength of the final product [38].

10.7.3 Energy storage

The powder may be used in energy storage devices such as batteries and supercapacitors as an electrode material. Ball milling generates powder with a wide surface area and tailored properties, which may improve the electrochemical performance of electrodes [39]. Furthermore, titanium-based electrodes may increase the material's stability, conductivity, and cyclability, allowing it to be used in a variety of energy storage applications.

10.7.4 Catalysis

The powder material has the potential to work as a catalyst in a variety of processes, such as carbon-carbon coupling, oxidation, and hydrogenation. Ball milling may enhance the catalytic activity and selectivity of a catalyst by increasing the surface area of the powder and modifying its physical properties. Furthermore, titanium-based catalysts may be less expensive and less harmful to the environment than conventional precious metal-based catalysts [40].

10.7.5 Coatings

Ball milling generates powder with fine particles and customised characteristics, which may increase durability, homogeneity, and coating adherence. Moreover, the titanium-based coatings have the ability to improve the material's mechanical and corrosion resistance properties, which makes them suitable for a wide range of industrial applications [41].

10.8 CHALLENGES

Ball milling of MCs into feedstock powder is a potential strategy to reduce waste and improve sustainability in the manufacturing sector. However, various problems must be overcome in order to assure the quality and uniformity of the feedstock powder. One of the most difficult issues in removing impurities from MCs is removal of oils, greases, and cutting fluids used throughout the machining process to lubricate the cutting tool and cool the workpiece [42]. These impurities may attach to the surface of the chips, altering the characteristics and sintering ability of the feedstock powder. These pollutants might potentially cause safety issues during the ball milling process. As a result, developing efficient cleaning procedures to remove these impurities from MCs prior to ball milling is critical. Achieving a consistent particle size distribution, phase composition, and morphology is another problem. The ball milling process is sensitive to milling settings and the milling environment, which may result in differences in the qualities of the feedstock powder. The milling duration, speed, ball-to-powder ratio, and milling environment may all have an impact on the particle size distribution, crystalline structure, and morphology of the powder [43]. As a result, optimising these parameters is critical in order to get a consistent and homogenous feedstock powder. Furthermore, the feedstock powder's phase composition is crucial in defining its qualities and uses. Depending on the milling parameters, the ball milling process may generate phase transformations such as amorphisation, phase mixing, and solid-state reactions [12]. As a consequence, characterising the phase composition of the feedstock powder using techniques such as XRD, SEM, and TEM is essential to ensuring its suitability for the intended application.

10.9 FUTURE SCOPE

In sustainable manufacturing, ball milling offers enormous potential for reusing MCs. Future research should concentrate on finding new techniques to remove impurities from MCs as well as enhancing the ball milling process to generate feedstock powder with the appropriate qualities. To increase confidence in the viability of this process, it is necessary to compare the in-process performance of this substrate to that of gas-atomised, plasma-atomised, and milled powder in terms of deposition efficiency and flowability and to study the dynamic mechanical properties, such as hardness and fatigue life. Additionally, there is a need for study into additional waste materials, such as building waste and electronic waste, that may be used as powder feedstock for powder metallurgy. Finally, the environmental and economic advantages of recycling MCs using ball milling over conventional production methods must be assessed.

REFERENCES

[1] F.H. Froes, B. Dutta, The additive manufacturing (AM) of titanium alloys, *Adv. Mater. Res.* 1019 (2014) 19–25.

[2] K.J. Martchek, The importance of recycling to the environmental profile of metal products *Fourth International Symposium on Recycling of Metals and Engineered Materials. Proceedings of the Minerals, Metals & Materials Society (TMS)* (2000) 19–28.

[3] B. Rotmann, B. Friedrich, C. Lochbichler, Challenges in titanium recycling - do we need a new specification for secondary alloys?, *Proc. - Eur. Metall. Conf. EMC 2011.* 4 (2011) 1465–1480.

[4] D. Paraskevas, K. Vanmeensel, J. Vleugels, W. Dewulf, Y. Deng, J.R. Duflou, Spark plasma sintering as a solid-state recycling technique: The case of aluminum alloy scrap consolidation, *Materials (Basel).* 7 (2014) 5664–5687.

[5] M.L. Hu, Z.S. Ji, X.Y. Chen, Q.D. Wang, W.J. Ding, Solid-state recycling of AZ91D magnesium alloy chips, *Trans. Nonferrous Met. Soc. China (English Ed).* 22 (2012) s68–s73.

[6] A.E. Tekkaya, M. Schikorra, D. Becker, D. Biermann, N. Hammer, K. Pantke, Hot profile extrusion of AA-6060 aluminum chips, *J. Mater. Process. Technol.* 209 (2009) 3343–3350.

[7] W.Z. Misiolek, M. Haase, N. Ben Khalifa, A.E. Tekkaya, M. Kleiner, High quality extrudates from aluminum chips by new billet compaction and deformation routes, *CIRP Ann. - Manuf. Technol.* 61 (2012) 239–242.

[8] O. D. Neikove, *Mechanical Alloying, Handbook of Non-Ferrous Metal Powders*, 2nd ed., 2009, Elsevier Ltd, Oxford, UK.

[9] S. Dhiman, R.S. Joshi, S. Singh, S.S. Gill, H. Singh, R. Kumar, V. Kumar, Recycling of Ti6Al4V machining swarf into additive manufacturing feedstock powder to realise sustainable recycling goals, *J. Clean. Prod.* 348 (2022) 131342.

[10] F. Erdemir, Study on particle size and X-ray peak area ratios in high energy ball milling and optimization of the milling parameters using response surface method, *Meas. J. Int. Meas. Confed.* 112 (2017) 53–60.

[11] G. Ingarao, R. Di Lorenzo, F. Micari, Sustainability issues in sheet metal forming processes: an overview, *J. Clean. Prod.* 19 (2011) 337–347.

[12] S. Dhiman, R.S. Joshi, S. Singh, S.S. Gill, H. Singh, R. Kumar, V. Kumar, A framework for effective and clean conversion of machining waste into metal powder feedstock for additive manufacturing, *Clean. Eng. Technol.* 4 (2021) 100151.

[13] H.C.S. Subasinghe, A.S. Ratnayake, General review of titanium ores in exploitation: present status and forecast, *Comun. Geol.* 109 (2022) 21–31.

[14] K. Mahmood, W.U.H. Syed, A.J. Pinkerton, Innovative reconsolidation of carbon steel machining swarf by laser metal deposition, *Opt. Lasers Eng.* 49 (2011) 240–247.

[15] R. Li, Y. Shi, Z. Wang, L. Wang, J. Liu, W. Jiang, Densification behavior of gas and water atomized 316L stainless steel powder during selective laser melting, *Appl. Surf. Sci.* 256 (2010) 4350–4356.

[16] A. Khaliq, M.A. Rhamdhani, G. Brooks, S. Masood, Metal extraction processes for electronic waste and existing industrial routes: a review and Australian perspective, *Resources.* 3 (2014) 152–179.

[17] J.M. Cullen, J.M. Allwood, Mapping the global flow of aluminum: from liquid aluminum to end-use goods, *Environ. Sci. Technol.* 47 (2013) 3057–3064.

[18] S.F. Sibley, Overview of flow studies for recycling metal commodities in the united states, *U.S. Geol. Surv. Circ.* 1196-AA (2011) 31.

[19] F. Bonciu, The European economy: from a linear to a circular economy, *Rom. J. Eur. Aff.* 14 (2014) 78–91.

[20] C.M. Lee, Y.H. Choi, J.H. Ha, W.S. Woo, Eco-friendly technology for recycling of cutting fluids and metal chips: a review, *Int. J. Precis. Eng. Manuf. - Green Technol.* 4 (2017) 457–468.

[21] I. Moravcik, A. Kubicek, L. Moravcikova-Gouvea, O. Adam, V. Kana, V. Pouchly, A. Zadera, I. Dlouhy, The origins of high-entropy alloy contamination induced by mechanical alloying and sintering, *Metals (Basel).* 10 (2020) 1–15.

[22] F. Khodabakhshi, A.P. Gerlich, Potentials and strategies of solid-state additive friction-stir manufacturing technology: a critical review, *J. Manuf. Process.* 36 (2018) 77–92.

[23] T. Fedina, J. Sundqvist, J. Powell, A.F.H. Kaplan, A comparative study of water and gas atomized low alloy steel powders for additive manufacturing, *Addit. Manuf.* 36 (2020) 101675.

[24] G.P. de León, V.E. Lamberti, R.D. Seals, T.M. Abu-Lebdeh, S.A. Hamoush, Gas atomization of molten metal: part I. Numerical modeling conception, *Am. J. Eng. Appl. Sci.* 9 (2016) 303–322.

[25] M. Kumar, X. Xiong, Z. Wan, Y. Sun, D.C.W. Tsang, J. Gupta, B. Gao, X. Cao, J. Tang, Y.S. Ok, Ball milling as a mechanochemical technology for fabrication of novel biochar nanomaterials, *Bioresour. Technol.* 312 (2020) 123613.

[26] M. Wei, B. Wang, M. Chen, H. Lyu, X. Lee, S. Wang, Z. Yu, X. Zhang, Recent advances in the treatment of contaminated soils by ball milling technology: classification, mechanisms, and applications, *J. Clean. Prod.* 340 (2022) 130821.

[27] M.S. Kamble, K. Bhosale, M. Mohite, S. Navale, Methods of preparation of nanoparticles, *Int. J. Adv. Res. Sci. Commun. Technol.* 7 (2022) 640–646.

[28] S. Hong, B. Kim, Effects of lifter bars on the ball motion and aluminum foil milling in tumbler ball mill, *Mater. Lett.* 57 (2002) 275–279.

[29] J.Y. Park, Y.M. Gu, S.Y. Park, E.T. Hwang, B. Sang, J. Chun, J.H. Lee, Two-stage continuous process for the extraction of silica from rice husk using attrition ball milling and alkaline leaching methods, *Sustainability.* 13 (2021) 7350.

[30] V. Bulgakov, S. Pascuzzi, S. Ivanovs, G. Kaletnik, V. Yanovich, sciencedirect angular oscillation model to predict the performance of a vibratory ball mill for the fine grinding of grain, *Biosyst. Eng.* 171 (2018) 155–164.

[31] J. Hao, W. Leong, E. Wong, K. William, An overview of powder granulometry on feedstock and part performance in the selective laser melting process, *Addit. Manuf.* 18 (2017) 228–255.

[32] S.N. Naik, S.M. Walley, The Hall-Petch and inverse Hall-Petch relations and the hardness of nanocrystalline metals, *J. Mater. Sci.* 55 (2020) 2661–2681.

[33] M. Ahmed, M. Pasha, W. Nan, M. Ghadiri, A simple method for assessing powder spreadability for additive manufacturing, *Powder Technol.* 367 (2020) 671–679.

[34] A.B. Spierings, M. Voegtlin, T. Bauer, K. Wegener, Powder flowability characterisation methodology for powder-bed-based metal additive manufacturing, *Prog. Addit. Manuf.* 1 (2016) 9–20.

[35] D. Dong, X. Huang, J. Cui, G. Li, H. Jiang, Effect of aspect ratio on the compaction characteristics and micromorphology of copper powders by magnetic pulse compaction, *Adv. Powder Technol.* 31 (2020) 4354–4364.

[36] S. Shrestha, K. Chou, Single track scanning experiment in laser powder bed fusion process, *Procedia Manuf.* 26 (2018) 857–864.

[37] V. Piotter, B. Zeep, P. Norajitra, R. Ruprecht, A.von der Weth, J. Hausselt, Development of a powder metallurgy process for tungsten components, *Fusion Eng. Des.* 83 (2008) 1517–1520.

[38] A.M. Beese, B.E. Carroll, Review of mechanical properties of Ti-6Al-4V made by laser-based additive manufacturing using powder feedstock, *Jom.* 68 (2016) 724–734.

[39] P. Thomas, C.W. Lai, M.R. Bin Johan, Recent developments in biomass-derived carbon as a potential sustainable material for super-capacitor-based energy storage and environmental applications, *J. Anal. Appl. Pyrolysis.* 140 (2019) 54–85.

[40] S. Dosta, M. Robotti, S. Garcia-Segura, E. Brillas, I.G. Cano, J.M. Guilemany, Influence of atmospheric plasma spraying on the solar photoelectro-catalytic properties of TiO_2 coatings, *Appl. Catal. B Environ.* 189 (2016) 151–159.

[41] W. Wang, C. Zhang, P. Xu, M. Yasir, L. Liu, Enhancement of oxidation and wear resistance of Fe-based amorphous coatings by surface modification of feedstock powders, 73 *Mater. Des* (2015) 35–41.

[42] G.S. Goindi, P. Sarkar, Dry machining: A step towards sustainable machining - challenges and future directions, *J. Clean. Prod.* 165 (2017) 1557–1571.

[43] F. Hadef, Synthesis and disordering of B2 TM-Al (TM = Fe, Ni, Co) intermetallic alloys by high energy ball milling: a review, *Powder Technol.* 311 (2017) 556–578.

Chapter 11

Additive manufacturing for Industry 4.0

Francis Luther King M, Robert Singh G, and Gopichand A
Swarnandhra College of Engineering and Technology

Srinivasan V
Annamalai University

11.1 INTRODUCTION

The fourth industrial revolution, also known as Industry 4.0, has hallmarks similar to the previous three in that it is marked by the introduction of new technologies and processes that have the potential to radically alter economies, occupations, and social structures. Automation, computerization, the Internet of Things (IoT), artificial intelligence (AI), and machine learning (ML) are permeating every aspect of modern life and industry. One area of Industry 4.0 seeing rapid growth is additive manufacturing (AM). They are no longer cutting-edge research; rather, 3D printing has entered a new age that will have far-reaching effects on the ways in which businesses operate. In recent years, the increasing capabilities of AM, together with the growth in accessible materials and the ecosystem of suppliers, have allowed for the low-cost production of a far broader variety of objects, from basic models of gaming figurines to pieces of vehicles or aeroplanes. With AM, businesses can tailor their offerings to meet the needs of individual customers and adapt rapidly to changes in consumer demand. This means it is expanding beyond niche uses like prototyping and making traditional equipment play a more essential role in production for a growing range of sectors [1]. In order to make components and items, manufacturing and crafting systems up until recent decades mainly depended on a few fundamental processes. These methods included sculpting, drilling, moulding, folding, merging or glueing numerous materials, or combining two or more of these methods. Drilling and sculpting involve progressively reducing material from the original input matter until the desired product is formed [2]. Business practises are expected to evolve gradually as a result of Industry 4.0, the fourth industrial revolution. It incorporates digital and mechanical processes such as analytics, robotics, AM, AI, advanced materials, natural language processing, high-performance computing, cognitive

DOI: 10.1201/9781003406488-11

technologies, and augmented reality (AR). Using tools such as 3D scanners, it takes data from the real world and digitizes it [3].

In the case of AM technology, it employs AI to facilitate digital information sharing and the conversion of digital data to physical form. In this field of technology, data is stored as 3D digital files, which are then utilized to create 3D physical items using a variety of AM techniques [4]. Industry 4.0 relies heavily on automation, cloud computing, the IoT, and other cutting-edge industrial technology. Industry 4.0 proposes a synergy between cyber and physical systems, with the end goal of erecting "smart factories" by rethinking the human workforce's function [5]. The primary objective is the development of a smart manufacturing system with an associated informational system (the smart supply chain) that engages in automated life-cycle management. By incorporating cutting-edge production technology, Industry 4.0 promises to revolutionize the industrial sector in the years to come. Through its advanced qualities of non-traditional mass customization, AM is renowned for being extremely time-saving and cost-effective for small-batch, complicated geometries of items, enabling firms to update and enhance their tools more often and swiftly [6]. Through an online survey administered to working professionals in AM-related firms in the UK, Europe, and the US, this article will discuss how AM tends to assist the demands of customers and markets in order to be more embraced by companies at some point [7]. There are many reasons why this cutting-edge technology is becoming increasingly popular, including the opportunity to differentiate oneself from competitors on the basis of environmental and social merits. New, cutting-edge offerings improve the company's flexibility, agility, and time-to-market. Rapid production is taking a giant leap forward with the help of AM.

The production of patient-specific implants and other high-quality medical equipment has also been significantly impacted by Industry 4.0. Detection of glucose levels and electronic-based testing are two examples of how this satisfies the need for a bespoke solution. Now, thanks to this change, doctors and surgeons may use their creativity to create a working model of a new medical treatment before committing to it for real. It's more likely that they'll be able to make a breakthrough that will aid in the research and development processes and eventually bring innovation to this field. AM is rising to the challenge posed by customization and the digital production system in the current iteration of Industry 4.0. So, it's important to investigate the ways in which AM may be integrated into Industry 4.0 to meet a wide range of future needs at a lower total cost and in less time. Understanding AM's potential in the future manufacturing system is deepened by this paper. However, putting AM to work in real-world settings is still difficult. There are numerous more aspects, both technological and social as well as legal, that must come together for this cutting-edge technology to be successful. Thus, numerous difficulties have developed with the introduction of this revolutionary technology, including a lack of knowledge and competence in AM, the quality of printed items, and the absence of standards and rules to its procedures.

11.2 INDUSTRIAL REVOLUTIONS

The transition from artisanal to factory manufacturing marks the beginning of the industrial revolution. It alters customs and trends in order to make people more at ease. As a result, new energy-efficient manufacturing methods were used to bring about these shifts. It also makes use of cutting-edge modes of transportation to enhance daily living. Before the industrial revolution, most people had to rely on horses or boats to go from one place to another. However, with the advent of new modes of transportation such as aeroplanes, vehicles, trains, and steamboats, the globe has become much more accessible.

Products that are in high demand across the world may be made using AM technologies, which make use of software and 3D scanners. More and more businesses in modern times are moving away from mass production and towards individualized production methods, which shorten the time it takes to make a certain product. Important improvements for contemporary industry may be attributed to AM. Many disciplines, including engineering, medicine, science, and others, make use of it. Accelerated innovation from AM is what drives the current trend towards completely tailored client needs. Figure 11.1 depicts the evolution of the industrial revolution.

11.2.1 The first industrial revolution (1.0)

Towards the close of the 18th century, between 1760 and 1840, the first industrial revolution began to take shape. It mechanized factories in an attempt to cut down on labour costs. By the time of the first industrial

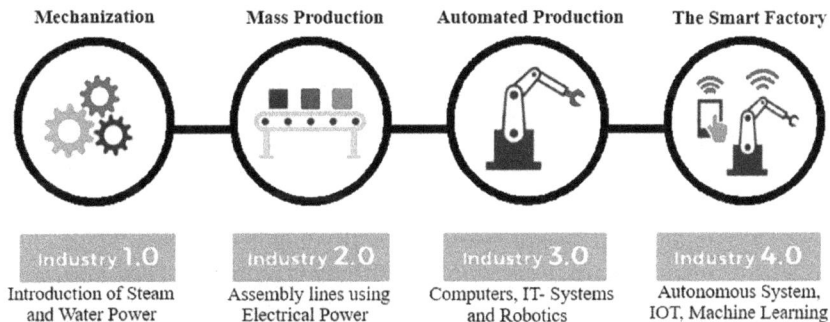

Mechanization	Mass Production	Automated Production	The Smart Factory
Industry 1.0	Industry 2.0	Industry 3.0	Industry 4.0
Introduction of Steam and Water Power	Assembly lines using Electrical Power	Computers, IT- Systems and Robotics	Autonomous System, IOT, Machine Learning

Figure 11.1 Evolution of industrial revolution.

revolution, manual production had been somewhat upgraded in the name of comfort thanks to the introduction of steam-powered engines. They did some fundamental modernization in the agricultural and textile sectors

11.2.2 The second industrial revolution (2.0)

There was an industrial revolution between 1870 and 1914 that brought about the widespread use of railways and telegraphs in manufacturing. Used in assembly lines, quality assurance labs, and inspection facilities. Though it sped up production, it offered relatively little leeway in terms of design or manufacture. Because of this, the mass manufacturing system could be implemented. Brought cutting-edge ideas from chemistry and related disciplines into production.

11.2.3 The third industrial revolution (3.0)

Once the 1950s rolled around, the stage was set for the third industrial revolution. Through the employment of robots and programmed flexible automation, this revolution brought about improvements in production quality, speed, and adaptability. Digital technologies are being integrated into manufacturing processes via the use of new machinery such as 3D printers, computer numerical control (CNC) machines, and computer-controlled robots. A variety of programs and hardware were supported by its infrastructure. It ushered in a period of rapid advancement in the fields of computing, communication, and IT. With this revolution, the mechanical world made the switch from analogue to digital. When applied to automation, this revolution ushered in a brand-new era of adaptable, customer-specific production methods that made use of a wide range of production techniques. The computer-integrated design and production system were improved, which is a huge boon to the procedure of creating new products.

11.2.4 The fourth industrial revolution (4.0)

The year 2020 is set as the target for the arrival of this fourth industrial revolution, which is expected to deliver an outstanding capacity to meet the industry's prospective demands. AM, the IoT, the industrial IoT, AI, and robotics are all crucial to this transformation. In order to significantly improve the industrial design, production, development, research, business model, and supply chain, tremendous creativity is required. This innovative approach to production will maximize the benefits of recently developed manufacturing methods and techniques. The goal is to build a smart factory capable of producing smart goods by integrating intelligent manufacturing equipment with information technology.

Digital solutions and cutting-edge technology, which many in the industry call "Industry 4.0," have made it feasible for the third and fourth industrial revolutions to converge. Some examples are:

- Automation and the IoT in Industry
- Large-Scale Data Collection
- To store data and programmes on a remote server
- Manufacturing techniques that use "additive" techniques (AM)
- Robotics with a high level of sophistication
- Realities that combine computer-generated imagery with the physical world are called augmented and virtual reality.

By connecting previously separate computer systems, these technologies are propelling the digital revolution of production. Many advantages become available to businesses as they adopt Industry 4.0, such as improved responsiveness, adaptability, personalization, and efficiency.

11.3 AM AFFORDABILITY PILLARS

In comparison to more conventional methods of production, AM has several advantages that make it the method of choice. The right approach to work is necessary to reap these rewards, however. On order to craft a successful strategy, you need to be well-versed in the advantages and how they pertain to you specifically. You need to educate yourself on the different technologies and the benefits they provide before diving into the available choices. The rapid prototyping technology of the 1990s pales in comparison to the AM technology of today. When compared to more conventional methods, AM offers three distinct advantages in terms of cost-effectiveness. These tenets will serve as the cornerstones of your AM plan.

11.3.1 The first supporting staple: Value stream mapping

AM typically uses just the materials needed for the component, which greatly reduces production waste. It also eliminates the need for the long lead times sometimes involved with making moulds, castings, and finished (direct-to-manufacture) goods in certain materials [8]. As part of a broader value stream mapping initiative, AM may be used to reduce lead times and waste. It can make the buy-to-fly ratio much closer to 1:1 than the typical 30:1. This is particularly important when producing limited quantities of unique things out of rare or expensive metals. The number of stages in a manufacturing process may be reduced thanks to AM, both in terms of tooling and direct fabrication, removing the need for human assembly to the point where the whole assembly of the correct item might be printed.

11.3.2 Functional designs cost-effectiveness

Changes to designs made late in the process, especially after tooling has been developed, may be very expensive. Those prices are usually excessive even if the product isn't seriously flawed. There's a possibility that the redesign won't be implemented. This implies that the product won't get 'nice-to-have' features or any other minor design tweaks. With its short lead times and simple flexibility to accommodate design modifications, AM technology may aid in the resolution of these issues. Instead of trying to avoid design changes, a company may focus on making a high-performance product without worrying about how those changes would affect production time and money. Companies who invest in producing goods of greater quality will likely see an increase in customer satisfaction, the number of consumers who are thrilled by those goods, the number of items that are returned, and their overall reputation.

11.3.3 Essential parts of cost-effectiveness

More fasteners, clips, glue, and other materials are needed for every component utilized in an assembly, which in turn increases the assembly's overall size, cost, and difficulty. In most cases, an assembly's performance improves as its number of parts decreases [9,10]. Many additively manufactured materials (AM) can now readily compete with the performance and requirements of conventional materials. Reduced assembly time and materials are the results of a smaller number of components.

11.4 INDUSTRY 4.0 IN A NUTSHELL

As a result of Industry 4.0, the conventional manufacturing sector is undergoing transformation. The term "Industry 4.0," sometimes known as the "Fourth Industrial Revolution," describes the convergence of three technological developments: ubiquitous connection, robust intelligence, and adaptable automation. With the advent of Industry 4.0, a cyber-physical system has been established via the merging of IT and OT. The advent of digital solutions and cutting-edge technology, often linked with the term "Industry 4.0," has allowed for this convergence. Digital transformation in manufacturing is being fuelled by these technologies, which allow for the seamless connection of formerly separate computer systems and activities throughout the whole value and supply chain. A company may gain increased agility, flexibility, and operational effectiveness by adopting Industry 4.0, digital manufacturing, and the interconnection that comes with it.

11.4.1 Industrial Internet of Things (IIoT)

The IoT is at the core of the fourth industrial revolution (IoT). IoT is shorthand for a network of physically linked objects that are digitally integrated to allow for communication and the exchange of data via the Internet.

Mobile phones, kitchen appliances, automobiles, and even whole structures might all be examples of "smart gadgets." The term "Internet of Things" refers to the network of interconnected devices, services, and sensors that connect everyday objects to the web. Thanks to technological advancements in recent decades, the Internet can now be expanded to a new level known as "smart objects," which forms the basis of an IoT vision. This novel pattern consists of endowing everyday objects with intelligence, allowing them to not only gather data and collaborate with their environment but also to be interconnected with other objects, communicate data, and run tests over the Internet. Several different concepts and definitions for the IoT [1] have emerged in response to the expanding interest in this area, which is generally recognized as one of the key drivers of Industry 4.0.

To establish a smart manufacturing environment, also known as a smart factory, the IoT refers to the connectivity of devices, vehicles, buildings, and other things equipped with electronics, software, sensors, actuators, and network connections to collect and exchange data. Additionally, "The Internet of Services (IoS)" idea applies a similar technique to services rather than physical assets, much as "The Internet of Things" does with "things." New opportunities will arise in the service industry as a result of the IoS concept, which lays the economic and technical groundwork for the development of business networks between service providers and customers [11].

There will be many benefits for users, producers, and companies as IoT spreads across industrial settings and value chains, which will have far-reaching effects on many sectors. Numerous innovative uses for the IoT are springing up around three central themes: i) process optimization, ii) resource optimization, and iii) the development of complex autonomous systems. The proliferation of IoT-based technologies will lead to an increase in the intelligence, dependability, and autonomy of everyday things, paving the way for the distribution of increasingly valuable goods and services [1]. However, the efficiency of Industry 4.0 is dependent on the preexisting network infrastructure, the intelligence, and the human expertise integrated into the system [12].

11.4.2 Big Data and analytics

Big Data is the term used to describe the massive amounts of data and their inherent complexity that are produced by IoT gadgets. This information is generated from a broad variety of resources, including cloud and corporate apps, websites, computers, sensors, cameras, and more, and it is sent through a wide variety of protocols.

Data from ERP, CRM, and MES system databases, as well as data from production equipment equipped with sensors, are just a few examples of the various forms of data that must be considered in the manufacturing business. Massive amounts of data are being created constantly by all aspects of our surroundings. Data is produced by every digital action and social media interaction. Information like that is sent by systems, sensors, and mobile devices. Multiple sources are contributing to the influx of big data at an unprecedented rate, with an equally staggering amount of data and a wide range of formats. Maximum processing capacity, analytical ability, and information management skills are required to get actionable value from big data [3]. The world around us is filled with a plethora of disparate data sets. It seems hard to make this environment very intelligent without appropriately employing data mining techniques. Depending on the automation system in place, data mining may be either supervised, unsupervised, or based on reinforcement learning. The precision of computer-assisted learning improves when it is implemented in a hierarchical structure with several levels. Automatic feature extraction through supervised or unsupervised learning in a hierarchical framework (ML) is the goal of ML [13].

11.4.3 Cloud systems

Applications such as remote services, colour management, and performance testing all make use of the cloud computing model. It has captured the imagination of the IT industry, and its importance in other facets of business is only going to increase. As infrastructure, data management, and capabilities continue to evolve, they will increasingly move to the cloud. Faster distribution, easier updates to newer versions, and more up-to-date performance models are all made possible by the cloud [14]. By "cloud computing," we mean the practise of relying on an external, remote server to house one's computer's operating system, programmes, and data. Efficiency is increased since all stakeholders, including input providers, workers, and end users, may see the same data at the same time [12]. With cloud computing, you may save money, streamline your infrastructure, increase your office space, protect your data, and get to it instantly. You may categorize the systems into four broad categories: i) the public cloud, ii) the private cloud, iii) the hybrid cloud (which combines the public and private clouds), and iv) the community cloud, which describes the sharing of any cloud-based service with a select group of businesses [15]. Big Data (either structured or unstructured) management solutions may be found in cloud computing environments. If you need to analyse a massive quantity of data but your local computer is struggling to keep up, a cloud-based solution is a far more practical and time-saving alternative. Therefore, it's essential that Industry 4.0 include data analysis and cloud-based infrastructure. The influence and pervasiveness of cloud-connected robots are growing rapidly [3]

11.4.4 Advanced robotics

Despite robotics' long history of service in the industrial sector, the advent of Industry 4.0 has revitalized the field. Recent technological developments have led to the development of a new generation of sophisticated robots that can carry out complex and nuanced jobs with ease. They are equipped with state-of-the-art software and sensors, allowing them to observe their surroundings, interpret that data, and take appropriate action, as well as work in tandem with and learn from people.

Collaborative robots ("Cobots"), which are built to operate in close proximity to humans while minimizing risk to both parties, are a rapidly growing field in robotics. As a consequence of technical developments, new products and methods of operation appear on the market every day. There is potential for flying cars, holographic television, and the implantation of hundreds of electronic devices inside the human body [16]. In the not-too-distant future, humanoid robots will be commonplace. Robots already have the ability to influence their surroundings, thanks to recent technological advancements. Teams of robots that work together to accomplish goals that have been previously established will become more commonplace as a result of advancements in AI [17].

Incorporating a collaborative robot into a manufacturing facility has many advantages for the business, including i) reducing the risk of injury and illness to employees by eliminating repetitive, non-ergonomic tasks; (ii) increasing productivity and product quality while decreasing costs; and (iii) making the company more competitive with respect to countries that offer lower wages. The advantages of robot use in a productive process are amplified when they are added to the work of an operator. On the first level, humans and robots do not operate together. In the end level [18], the man and the robot work side by side in the same office.

11.4.5 Additive manufacturing

AM, often known as 3D printing, is a crucial technology propelling Industry 4.0 alongside robots and cognitive systems. AM is the process of making physical objects from digital 3D models using a 3D printer. In the framework of the fourth industrial revolution, 3D printing is becoming an important tool for digital production. Rapid prototyping has evolved into AM, which has broad applications in production, from tooling to mass customization. It facilitates a form of distributed manufacturing in which components are kept as digital blueprints in online warehouses and manufactured at the point of consumption. A decentralized manufacturing system eliminates the need for lengthy and expensive shipping of materials, and it also makes inventory management easier by keeping track of digital information rather than physical components.

11.4.6 Virtual replicas or "Digital Twins"

A digital twin has the potential to greatly improve the efficiency and upkeep of manufacturing infrastructure. According to Gartner, a leading market research company, by 2021, half of the world's largest industrial enterprises will use digital twins to track and manage their physical assets and operational procedures. In order to better understand, analyse, and optimize their operations, businesses are increasingly turning to what is known as "digital twins," or digital representations of physical products, machines, processes, or systems that can be run in real time. There are certain similarities between digital twins and engineering simulation, although the latter misses the point. A digital twin, in contrast to engineering simulations, is an online simulation that uses data from sensors embedded in a machine or other equipment. Since IIoT devices transmit data in near-real time, a digital twin may continually gather this data, staying true to the original throughout the product's or system's lifetime. Due to this, the digital twin can foresee problems and take preventative action. A digital twin may be used by an operator to do things like figure out why a component isn't working or estimate how long a product will last. This constant simulation aids in the enhancement of product designs and the maintenance of equipment uptime. Digital twins have been widely used in high-stakes industries including aircraft, heavy equipment, and automotive. These days, the idea of digital twinning is being applied to a wider range of fields because of developments in computers, ML, and sensors.

11.4.7 Augmented reality

AR has been widely adopted in consumer contexts, but the industrial sector is just now starting to investigate its potential. There is a lot of unrealized potential in technology, such as in the realms of assembly and industrial equipment maintenance. Virtual visuals or data are superimposed over a real-world item in AR, creating a seamless connection between the digital and real worlds. Smart phones, tablets, and AR glasses are used for this purpose. Take the use of AR glasses by a surgeon during a procedure as an example in the medical field. An individual's nerves, main blood vessels, and ducts might be highlighted in colour by MRI or CT scan data that could be superimposed on the individual via the glasses. The surgeon is aided in determining the least hazardous entry point into the target area, thereby reducing the likelihood of problems and maximizing the accuracy of the incision. In the realm of product design, development, and manufacturing, a virtual model fed with information from big data and cloud technologies may help to assess any and all potential outcomes. Widespread usage of simulation in business models allows for the emulation of the real-world working environment in a virtual ecosystem, making use of the available real-time data. Before making changes in the real world, individuals may test and optimize processes in a virtual environment to reduce business turnover, risk, setup time, and improve quality control for future processes and services [12]. Computer-generated images are superimposed over real-world views in a process known as simulation and AR. This technology is crucial to the functioning of Industry 4.0 applications. This cutting-edge technology, essential to the ongoing industrial transformation, was developed via the merger of the operational and simulation sectors [11]. There are several benefits to using these methods, particularly in the context of product development and production. The development of smart manufacturing capabilities is facilitated by AR, one of the cutting-edge technologies featured in the Industry 4.0 movement [18].

11.4.8 Enterprise resource planning and business intelligence

The term "enterprise resource planning" (ERP) is used to describe computerized information systems that aim to coordinate and maximize the use of an enterprise's various resources. ERP software is a system that helps a company centralize data and procedures that are carried out in different departments and locations (suppliers, production, stock, and sales). In order to begin predicting and extracting information that may be used in many departments, ERP systems are able to provide an integrated strategy to data utilization [11]. Industry 4.0, MES, the cloud, and ERP all work together, and big data is an integral part of all of these processes. All design processes, as well as the whole customer experience, must be fully compatible with the Industry 4.0 paradigm. Important to this structure is the use of ERP software [17]. The concept of Industry 4.0 requires standards for interoperability and cooperation. Feedback from the end-user is essential, as is the provision of instant value to all stakeholders. Personalization requires sophisticated network systems, which can only be achieved through careful design [12]. Telecom providers could be able to gauge customer happiness by analysing network performance over oscillations and using preventative scenario planning. These functions are made possible by an ERP system with a well-designed framework. The benefits of ERP systems make them a useful tool for implementing Industry 4.0. i) Customers can easily track their orders online and get the information they need in a timely manner [11]; ii) ERP systems can provide sales and purchasing transparency; iii) ERP data can be used by mobile applications to communicate; iv) Optimal resource utilization can be achieved under varying job descriptions; v) Real-time data analysis can be performed and allow for early detection.

11.4.9 Smart virtual product development system

The smart virtual product development (SVPD) system is a decision-support tool for product development that stores, applies, and disseminates Set of Experiences (SOEs) representing the accumulated wisdom from prior decision-making events. It was developed to fill the need for digitally capturing product design, production, and inspection information in smart manufacturing. Industry 4.0 principles demand this kind of improvement in both product quality and development time, and this is exactly what will happen [19].

11.4.10 Machine-to-machine communication

The term "machine-to-machine" (M2M) is used to describe the technology that enables wireless or wired connections for two machines to communicate directly with one another. Industrial instrumentation and private communications are two examples of the types of M2M communication that are possible [20]. The IoT and M2M communication are also vital to Industry 4.0. M2M communication is a system that enables machines to exchange data with one another via any available connection, whether wired or wireless. Industrial instrumentation and private networks are both examples of M2M communication. The IoT and M2M communication are also vital to Industry 4.0. The applications are designed to benefit businesses by increasing their value via the creation of new income streams and the elimination of wasteful expenditures [21] lays out the requirements for M2M operations, which include the following: (i) Remote Service and Asset Information Management providing, which facilitates information federation and lifecycle support; (ii) Vehicles that are connected to one another and one another's infrastructure, resulting in new social connections and dynamics like Retail, the supply chain, and related components, (iii) smart vending [3]. Developing a language for communication between machines and people, as well as between humans in various places, is only one of the many challenges brought up by the M2M vision [18]. Other concerns include creating smart environments, smart architecture, and a smart grid with wireless sensors.

11.5 INTELLIGENT MANUFACTURING OR SMART FACTORIES

Smart factories, also known as smart manufacturing, are a style of manufacturing that uses automation and digitalization in place of more conventional techniques in order to enhance the design, production, and engagement with their products. To achieve completely flexible production at the fastest possible speed [18]. It seeks to use cutting-edge information and manufacturing technology. Although the phrase "smart factory" has been used interchangeably with "dark factory," "lights-off factory," and "unmanned factory," the technology is designed to require very little human oversight. The person often engages with these systems throughout the process of attempting to solve an issue. Light-off (dark) factories, also known as unmanned factories, are modern manufacturing facilities that rely only on automated processes and autonomous machinery to run their operations [11,16]. The fact that dark factories can function without human labour is one of their most recognizable features. There's not enough time for the raw materials to be brought in, processed, and sent out of an unmanned facility. That is to say, all manufacturing in these facilities is done by robots [21]. Fourth industrial revolution requirements for smart factories' traits and practises seem like a given at this point. Not to mention these procedures, which will play a crucial role in the development of our manufacturing capabilities. Integrating additional components, such as big data, CPS, cloud, IoT, and M2M [11], is also crucial for creating a smart factory operating under Industry 4.0.

The construction of smart factories is contingent on a number of factors, including the cost and reliability of energy sources, the productivity of workers, and the accessibility of the appropriate technical infrastructure. These factories, however, will have a deleterious effect on current jobs and lead to higher unemployment rates [22].

11.6 ADDITIVE MANUFACTURING WORKFLOW

As a first step in an AM process, a computer-aided design (CAD) model of the artefact is created, detailing the geometry of the component to be manufactured. The process of making it a reality is similar to that of making any other physical object. Many researchers have tried to break AM down into its component parts [23]. They break this down into four stages in their consideration of the subject. Using either solid-modelling CAD software or by 3D scanning an actual object and transferring the data into the CAD environment, a digital model of the artefact is developed during the design process. The next step in the process planning phase is to utilize the data from the 3D model to set up the AM machine. Manufacturing occurs layer by layer as the AM system constructs the physical artefact. Post-processing requirements, such as heat treatment, are met during the finishing process, which also includes removing the support material. The verified item is next reviewed and validated by being compared to the original CAD model. Since the four-stage description doesn't cover everything, Kim et al. [24] divided it into eight stages linked by seven actions.

The manufacturing file should include all the production criteria needed to make a successful print. AM file standards seem to be mostly focused on standardizing the specification of CAD geometry, such as with the STL file improvement efforts of AMF and 3MF [25]. Tolerance data, item packing information, machine routes, support structures, slice structures, and object orientation are also necessary for a successful build in addition to the CAD geometry [26]. Technology independence, simplicity, scalability, and future compatibility are all features that should be taken into account when designing a stable file format [27, 26]. Geometrical data, for instance, should be process-independent, but tool paths and other details should be [28].

This indicates that data is dependent on the availability of certain equipment or devices. Data such as the order and timing of deposition routes, which affect the overall quality and composition of the final product with respect to stress and microstructure [23,24], is especially crucial for laser- and electron-beam AM processes. It is possible, following the slicing process, to move both the original geometry information (the native CAD file) and the tessellated features. The process of slicing is used to transform a 3D CAD model into machine-specific instructions, or G-code. Since there is currently no unified system in place for sharing information throughout the various phases of AM manufacturing, Four new file formats, in addition to the already-existing AMF format, were suggested by Nasser et al. [23,24]. It was proposed that the "slices" be documented in an "AMSF" format, with extensions for storing information on route planning and process parameters in "AMPF," sensor data and a qualification record in "AMQF," and final verification and validation results in "AMVF."

11.7 ADDITIVE MANUFACTURING DATA FORMATS

The STL file format has been the standard for sharing AM data models in recent years. STL is a computer format used to store data about tessellated surfaces. Although still widely used, the STL format has a number of drawbacks that make it less than ideal for use with modern AM machines that may use multiple nozzles and functionally graded materials. It is computationally and procedurally costly to protect against and repair STL files, including duplicate information and geometrical faults such as missing, overlapping, and degenerate facets. Another limitation of the STL format is its inability to save data on a model's materials, textures, colours, dimensions, or geometry [29]. The STL surface triangle information model does not include built-in capabilities for storing manifold surface information. Multiple proprietary options for storing 3D model data have been presented, and work is ongoing to develop standards that would fully satisfy the AM design criteria.

The AMF format is particularly noteworthy since it has been established and approved as an official ISO/ASTM standard for AM. AM uses the 3MF file format to help bridge the gap between hardware and software systems. Both of these formats are text-based and can be read and understood by people since they are based on the extensible markup language (XML), as stated by the open XML specification [30]. To organize the vast quantities of design and geometric data necessary for AM processes, the AMF and 3MF are actively establishing open standards. Whether or not AM hardware and software developers will refrain from disseminating their own controlled, closed-source formats or will instead adopt a single industry-wide framework capable of managing the intricacies of different AM processes and machines is yet to be seen. When it comes to standardizing file formats for

AM, there is an option in the form of the STEP and STEP-NC machining standards. The STEP criteria diverge from the norms in both product categories and printed forms. The shape and materials of the product, as well as any other necessary details, are described. STEP-NC, on the other hand, might be seen as a more comprehensive update to the original STEP standard. As such, a STEP-compliant product model may include the whole product lifecycle. Organizations such as the International Organization for Standardization and the American Society for Testing and Materials are collaborating to create universal product and manufacturing data standards. The current intent of both standards is to provide additions that may improve AM infrastructure and services. STEP contains data for geometric and tessellated models combined, unlike STL, AMF, and 3MF. STEP-NC aspires higher than STEP by adding more processing data, but it does so by eliminating the tessellated model in favour of a pure geometric one.

11.8 IMPACT OF AM IN BUSINESS MODELS

There seems to be limitless potential and long-term benefits for businesses that use AM technology during the fourth industrial revolution [31]. In the framework of Industry 4.0, AM is at the centre of several strategies and novel business models designed to advertise solutions and technologically improved alternatives to lessen the product's total environmental effect throughout its entire lifespan. New environmental perspectives can be developed as a result of the organizational and economic shifts taking place in companies' business models. This is especially true in regards to the decreased consumption of raw materials and transportation as a result of the growth of local cycles and the contraction of supply chains. Businesses will need to re-evaluate their operations and strategies in light of the effects of adopting AM. Consequently, AM promotes several options for sustainability via repair, refurbishment, and remanufacturing, which prolong the lives of goods and, by completing the loop, contribute to the sustainability of the system.

11.9 AM BUSINESS MODELS

AM is often called "3D printing" these days, but it is important to note that "3D printing" is not a technical term and is often used interchangeably with "AM." Until recently, the term "3D printing" was often used to describe machines that were cheap and/or had limited capabilities. In-depth explanations of the model's constituent parts and descriptions of the impact of additive technologies on those parts may be found in the specialized literature that proposes and discusses alternative business models [32]. Additive printing technologies and the new business paradigm that they've ushered in have also received a lot of attention [33]. Through an examination of the evolving technological landscape and the proliferation of AM, three distinct business model evolution paths have emerged. Models based on prototyping, close cooperation with the client, or producing customized customer solutions, as well as models based on core capabilities, make up the first, startup-focused orientation. Client collaboration and satisfaction are hallmarks of these strategies. The second trend involves a set of models predicated on mass production and designed to bring about the widespread implementation of technologically advanced goods and initiatives. The third route is based on the business model's value proposition, in the form of a solution proposal that may be used to determine what the business model provides from a scaling standpoint. The concept's creators recommend integrating AM's Industry 4.0 technology, aspects of business models and value propositions, and elements bringing the Scalability 4.0 impact into a company's business model for maximum efficiency.

11.9.1 Startup Business models (BM)

Over the last decade, 3D printing enterprises have proliferated. However, they are generally FDM machines without sophisticated gadgets [34]. They make cheap, weak models. Due to constraints, such goods cannot be utilized as automobile components or orthopaedic prostheses. Thus, microenterprises emphasize bespoke services. Their business concept is to make basic accessories or spare parts with a unit need (e.g., replacing an expired component with a certain shape) [35]. This business strategy usually requires an Internet shop, basic spatial design abilities, and a simple 3D printer to manufacture such an element fast, readily, and inexpensively. Competition and little innovation make this business model difficult to establish. These firms commonly supply consumables or replacement parts for this device.

Filaments for low-cost 3D printers are another business field. An intriguing business concept is partnering with a plastics manufacturer to purchase completed goods. Then offer it under your brand in your own cartons. This firm concentrates on sales, order processing, marketing, and mix development (which may be outsourced). 3D printer sales agents have several websites.

Advanced 3D printers should be considered while considering AM applications in different sectors. They enable complicated product geometry and mechanical qualities [36]. These machines process metal, polymer, and ceramic powders (e.g., Selective Laser Sintering (SLS), Selective Laser Melting (SLM), Electronic beam Melting (EBM), and 3 Dimensional Printing (3DP)) or resins (e.g., Stereolithography (SLA)). Automotive, aviation, and medical utilize these printers effectively [37]. This technology requires high AM design skills Design for Additive Manufacturing (DfAM) Design services and educating engineers to become experts are two business strategies in this sector. Printelize is an example in the first scenario.

Printelize Marketplace automates and optimizes 3D printing supplier discovery, appraisal, and sale [14]. 3D printing commissioners meet order takers here. This technology streamlines offerings, sales, and communication, making 3D printing simpler. 3D printing design is the most sought-after service talent. The capacity to develop spatial models using material processing technical expertise is what matters. Specialists are required to account for material shrinkage during 3D printing, optimize geometry for printing, or prepare it to avoid post-processing issues [34].

Understanding Industry 4.0, which includes AM, will help business models incorporate these technologies for mass manufacturing. AM replaces traditional manufacturing processes as the sector grows. Their versatility ensures efficient cost recovery and product customization. Small batches increase quality, simplify planning, and stimulate ongoing development. 3D printing lets you insert structural features into the project structure and mix materials with diverse qualities [38]. Manufacturers might restrict component production, reducing assembly time [39]. Also consider the requirement for less storage (broken things may be produced quickly without storing them). Optimization software allows firms to tailor output and enhance goods at a low cost without interrupting assembly line expansion or tooling changes. Medical and pharmaceutical business models may be changed. The FabRx machine can additively manufacture drug-dosed tablets (called printlets). This 3D printing method can also combine many medications into one tablet for individuals who need to take multiple tablets (children or the elderly) [40]. This approach is useful when mass manufacturing is unprofitable owing to low demand or high storage costs. Hospitals, pharmacies, and pharmaceutical businesses may benefit.

Personalization, product optimization, manufacturing efficiency (including production tools), and new business models make AM advances amazingly fast—2 years from concept to market in 2 months [41].

11.9.2 Mass AM BM

Reasons for the emergence of new business models include the advent of cutting-edge industrial technology. Frequently, they are iterations on pre-existing business models or the impetus for whole new ones to arise. The following is a list of recommended business models that are suitable for use with AM solutions or that have emerged as a direct consequence of the advent of AM.

Mass customization: It's the utmost in product customization since one-of-a-kind products are made to suit the tastes and preferences of each unique customer for a fraction of the price of mass-produced goods, all owing to the efficiency and accuracy of digital technology.

Mass segmentation: It severely restricts variety, offering only a few dozen versions of a product; it works well for highly segmented markets, such as components designed specifically for popular B2B products; each version serves a single segment and differs sufficiently from the others that conventional manufacturers would need costly new machine tools to make all of them; thus, additive companies can make them at a lower cost; production is in batches rather than continuous; and it works well for highly differentiated products.

Mass complexity: It takes advantage of AM's ability to produce products with intricate designs that conventional manufacturing cannot achieve, as well as to produce unusual shapes and embed sensors and other elements; this ability reduces production costs while improving product reliability; and with the advent of new design software, AM can now restructure materials at the micro level to improve properties such as porosity, strength, durability, elasticity, and rigidity.

Mass standardization: As 3D printers become more efficient, they may become competitive for making standardized products even when they do not save on direct costs; AM reduces a lengthy and risky supply chain, expensive capital equipment, the elaborate assembly of parts, and high inventory or transportation costs; and demonstrates that high-volume standard products can be churned out at a low cost under certain conditions.

Mass variety: Customers with strong and varied tastes are targeted; producers may forego gathering personal data; a large range of possibilities can be offered at reasonable rates; and goods are unique.

Theory of value proportion BM: Providing a value proposition to the consumer is one of the primary goals of a successful company model. In business model theory, worth may be understood in several ways. Five value areas—Value Proposition, Value Creation, Value Delivery, Value Capture, and Value Communication—are proposed by Osterwalder and Pigneur to describe the sectors of constructing a company model. The Value Proposition is a fundamental part of every successful company plan. The potential for AM and other Industry 4.0 technologies to be integrated into specific construction value domains is intriguing. In the long run, the traditional business model and Industry 4.0 technology may have something in common with the Scalability 4.0 idea. The notion of scalability will allow for the generation of the consequences of large-scale manufacturing, international commerce, and mass customization. What was previously unattainable owing to constraints in manufacturing technologies, transport technologies, and sales technologies due to, among other things, the cost of technological change, the cost of transport, and the cost of customizing individual offers is now feasible. Scalability 4.0 includes, in the authors' view, several characteristics: economies of scale and experience curve; network impact achieved within the network of suppliers, manufacturers, and consumers; long-tail effect; Big Data effect (advantage in data processing and analysis) [42]. As a result of integrating these three components:

Industry 4.0: Autonomous Robots, System Integration, IoT, Cyber Security, Cloud Computing, AM, AR, Big Data, Other; Value Proposition, Value Creation, Value Delivery, Value Capture, Value Communication, Other;

Classic value: The effects of mass production, mass trade, and the mass provision of customized services; new opportunities;

Scalability 4.0 concept: Because of advancements in technology, parts of the business model may be created. Because of it, it's possible to increase the worth of the building at that particular spot. Potentially affecting the scalability of a firm in the future. A synthesis of these ideas is shown in the figure. Choosing Industry 4.0 technology as a starting point for value creation is essential. Because of this, the value creation landscape shifts based on the specifics of the business model's components.

11.10 BENEFITS OF AM IN INDUSTRY 4.0

A fully automated, self-correcting, and optimally performing factory is the goal of Industry 4.0. A smart factory is one that can improve its own efficiency and safety measures, as well as those of its employees.

An astounding assessment of smart manufacturing's economic advantages has been given by the US Smart Manufacturing Leadership Coalition, a nationwide partnership of manufacturing businesses, research institutions, and government laboratories (Industry 4.0: The Benefits of Smart Manufacturing):

- There was a decrease in workplace accidents by 25%.
- Increased efficiency in the use of energy by 25%.
- Reducing water use by 40% and increasing total operational efficiency by 10%.
- The time it takes to bring a product to market has decreased by 10%.

11.10.1 Less waste and scrap

In contrast to conventional production methods, which include a series of subtractive steps, AM is a forward-looking process. AM involves the deliberate addition of material throughout the production process, as opposed to subtractive methods such as milling and turning. This results in less material and energy waste.

11.10.2 Shortens development cycles and lowers costs

The prototyping process is sped up, simplified, and less expensive with the help of AM. Milling is only one example of an expensive manufacturing technology that requires a lot of setup and raw materials. It's significantly easier to mass-create, test, and tweak a prototype because of the reduced time and money required. Moreover, it offers almost instantaneous evidence of the enhancements performed. To be successful, AM calls for constant and efficient communication between machines, gadgets, and robots. Only through the complete digitalization of production processes can this be achievable. To keep up with the demands of Industry 4.0, businesses are increasing their spending on digital and IoT infrastructure.

11.10.3 Synthesize the assembly part into a single part

AM in Industry 4.0 also streamlines the production process, especially product assembly. Complex and requiring numerous processes, traditional components are seldom used. This increases the time and effort needed to produce the components, as well as the cost of the materials used. However, with AM, you may print the whole group on a single sheet.

11.11 CONCLUSION

The purpose of this chapter is to explore and highlight many aspects of AMAM technology for a better understanding of this novel technology's role in the context of Industry 4.0. Because of its adaptability and promising future, AM is a vital forerunner in driving the fourth industrial revolution. Despite being a young field, AM has been gaining traction as a result of the many advantages it offers businesses. This chapter looked at AM's function in Industry 4.0 by analysing its effects and how they connect with the business models of AM companies, as well as by revealing the barriers to AM's widespread use in industrial settings. These aspects include customer satisfaction, impact on business models, and business sustainability, as well as the challenges that AM implementation may face in the context of Industry 4.0. This study provides an overview of current business models used in AM-based industries. Business models are broken down into three distinct categories by the authors: the startup method (company offering to market), mass manufacturing (market requirements to company), and value proposition (company and market cooperation). The three methods demonstrate the plethora of options available to businesses when implementing new technology in their operations. In such a scenario, new avenues for making sales and providing additional services open up. There are shared and unique characteristics among the various techniques.

REFERENCES

[1] Pereira, A. C., & Romero, F. (2017). A review of the meanings and the implications of the industry 4.0 concept. *Procedia Manufacturing*, 13, 1206–1214.
[2] Bae, E. J., Jeong, I. D., Kim, W. C., & Kim, J. H. (2017). A comparative study of additive and subtractive manufacturing for dental restorations. *The Journal of prosthetic dentistry*, 118(2), 187–193.

[3] Gunasekaran, A., Subramanian, N., & Ngai, W. T. E. (2019). Quality management in the 21st century enterprises: research pathway towards Industry 4.0. *International Journal of Production Economics*, 207, 125–129.

[3] Luthra, S., & Mangla, S. K. (2018). Evaluating challenges to industry 4.0 initiatives for supply chain sustainability in emerging economies. *Process Safety and Environmental Protection*, 117, 168–179.

[4] Haleem, A., & Javaid, M. (2018). Role of CT and MRI in the design and development of orthopaedic model using additive manufacturing. *Journal of clinical Orthopaedics and Trauma*, 9(3), 213–217.

[5] Gürdür, D., El-Khoury, J., Seceleanu, T., & Lednicki, L. (2016). Making interoperability visible: data visualization of cyber-physical systems development tool chains. *Journal of Industrial Information Integration*, 4, 26–34.

[6] Rodriguez-Conde, I., & Campos, C. (2020). Towards customer-centric additive manufacturing: making human-centered 3D design tools through a hand-held-based multi-touch user interface. *Sensors*, 20(15), 4255.

[7] Klahn, C., Fontana, F., Leutenecker-Twelsiek, B., & Meboldt, M. (2020). Mapping value clusters of additive manufacturing on design strategies to support part identification and selection. *Rapid Prototyping Journal*, 26(10), 1797–1807.

[8] Prashar, G., & Vasudev, H. (2021). A comprehensive review on sustainable cold spray additive manufacturing: state of the art, challenges and future challenges. *Journal of Cleaner Production*, 310, 127606.

[9] Manogharan, G., Wysk, R., Harrysson, O., & Aman, R. (2015). AIMS-a metal additive-hybrid manufacturing system: system architecture and attributes. *Procedia Manufacturing*, 1, 273–286.

[10] Poomathi, N., Singh, S., Prakash, C., Subramanian, A., Sahay, R., Cinappan, A., & Ramakrishna, S. (2020). 3D printing in tissue engineering: a state of the art review of technologies and biomaterials. *Rapid Prototyping Journal*, 26(7), 1313–1334.

[11] Oztemel, E., & Gursev, S. (2020). Literature review of Industry 4.0 and related technologies. *Journal of Intelligent Manufacturing*, 31, 127–182.

[12] Ghadge, A., Er Kara, M., Moradlou, H., & Goswami, M. (2020). The impact of Industry 4.0 implementation on supply chains. *Journal of Manufacturing Technology Management*, 31(4), 669–686.

[13] Sunhare, P., Chowdhary, R. R., & Chattopadhyay, M. K. (2022). Internet of things and data mining: an application oriented survey. *Journal of King Saud University-Computer and Information Sciences*, 34(6), 3569–3590.

[14] Youssef Abdelmajied, F. (2022). Industry 4.0 and Its Implications: Concept, Opportunities, and Future Directions. *Supply Chain - Recent Advances and New Perspectives in the Industry 4.0 Era*. Intechopen. https://doi.org/10.5772/intechopen.102520.

[15] Culot, G., Nassimbeni, G., Orzes, G., & Sartor, M. (2020). Behind the definition of industry 4.0: analysis and open questions. *International Journal of Production Economics*, 226, 107617.

[16] Mohammed, A., & Wang, L. (2018). Brainwaves driven human-robot collaborative assembly. *CIRP Annals*, 67(1), 13–16.

[17] Wang, L., Liu, S., Cooper, C., Wang, X. V., & Gao, R. X. (2021). Function block-based human-robot collaborative assembly driven by brainwaves. *CIRP Annals*, 70(1), 5–8.

[18] Tay, S. I., Lee, T. C., Hamid, N. Z. A., & Ahmad, A. N. A. (2018). An overview of industry 4.0: definition, components, and government initiatives. *Journal of Advanced Research in Dynamical and Control Systems*, 10(14), 1379–1387.

[19] Ahmed, M. B., Sanin, C., & Szczerbicki, E. (2019). Smart virtual product development (SVPD) to enhance product manufacturing in industry 4.0. *Procedia Computer Science*, 159, 2232–2239.

[20] Chen, M. (2013). Towards smart city: M2M communications with software agent intelligence. *Multimedia Tools and Applications*, 67, 167–178.

[21] Moreira, F., Alves, A. C., & Sousa, R. M. (2010). Towards eco-efficient lean production systems. In *Balanced Automation Systems for Future Manufacturing Networks: 9th IFIP WG 5.5 International Conference, BASYS 2010*, Valencia, Spain, July 21–23, 2010. Proceedings (pp. 100-108). Springer Berlin Heidelberg.

[22] Ojra, A. (2019). Revisiting Industry 4.0: A new definition. In Arai, K., Kapoor, S., & Bhatia, R. (eds.), *Intelligent Computing. SAI 2018. Advances in Intelligent Systems and Computing*, vol 858. Springer, Cham. https://doi.org/10.1007/978-3-030-01174-1_88.

[23] Nassar, A. R., & Reutzel, E. W. (2013). A proposed digital thread for additive manufacturing. In *2013 International Solid Freeform Fabrication Symposium*. University of Texas at Austin.

[24] Kim, D. B., Witherell, P., Lipman, R., & Feng, S. C. (2015). Streamlining the additive manufacturing digital spectrum: a systems approach. *Additive Manufacturing*, 5, 20–30.

[25] Xiao, J., Anwer, N., Durupt, A., Le Duigou, J., & Eynard, B. (2017). Standardisation focus on process planning and operations management for additive manufacturing. In *Advances on Mechanics, Design Engineering and Manufacturing: Proceedings of the International Joint Conference on Mechanics, Design Engineering & Advanced Manufacturing (JCM 2016)*, 14–16 September, 2016, Catania, Italy (pp. 223–232). Springer International Publishing.

[26] Pratt, M. J., Bhatt, A. D., Dutta, D., Lyons, K. W., Patil, L., & Sriram, R. D. (2002). Progress towards an international standard for data transfer in rapid prototyping and layered manufacturing. *Computer-Aided Design*, 34(14), 1111–1121.

[27] Hiller, J. D., & Lipson, H. (2009, September). STL 2.0: A proposal for a universal multi-material additive manufacturing file format. In *2009 International Solid Freeform Fabrication Symposium*. University of Texas Libraries.

[28] Lipman, R. R., & McFarlane, J. S. (2015). Exploring model-based engineering concepts for additive manufacturing. In *2015 International Solid Freeform Fabrication Symposium*. University of Texas Libraries.

[29] Chua, C. K., & Leong, K. F. (2014). *3D Printing and Additive Manufacturing: Principles and Applications (with Companion Media Pack)-of Rapid Prototyping*. World Scientific Publishing Company, Singapore.

[30] Bray, T., Paoli, J., Sperberg-McQueen, C. M., Maler, E., Yergeau, F., & Cowan, J. (1998). Extensible markup language (XML) 1.0.

[31] Huang, S. H., Liu, P., Mokasdar, A., & Hou, L. (2013). Additive manufacturing and its societal impact: a literature review. *The International Journal of Advanced Manufacturing technology*, 67, 1191–1203.

[32] Öberg, C., Shams, T., & Asnafi, N. (2018). Additive manufacturing and business models: current knowledge and missing perspectives. *Technology Innovation Management Review*, 8(6) 15–33.

[33] Savolainen, J., & Collan, M. (2020). How additive manufacturing technology changes business models?-review of literature. *Additive Manufacturing*, 32, 101070.

[34] Trzaska, R., & Mazgajczyk, E. (2020). Industry 4.0: overview of business models in additive manufacturing. *Education Excellence and Innovation Management: A*, 2025, 18453–18464.

[35] Geissbauer, R., Wunderlin, J., & Lehr, J. (2017). *The Future of Spare Parts is 3D: A Look at the Challenges and Opportunities of 3D Printing*. PwC Strategy.

[36] Davim, J. P. (Ed.). (2020). *Additive and Subtractive Manufacturing: Emergent Technologies* (Vol. 4). Walter de Gruyter GmbH & Co KG, Portugal.

[37] Tofail, S. A., Koumoulos, E. P., Bandyopadhyay, A., Bose, S., O'Donoghue, L., & Charitidis, C. (2018). Additive manufacturing: scientific and technological challenges, market uptake and opportunities. *Materials Today*, 21(1), 22–37.

[38] Yakout, M., & Elbestawi, M. A. (2017, May). Additive manufacturing of composite materials: An overview. In *Proceedings of the 6th International Conference on Virtual Machining Process Technology (VMPT)*, Montréal, QC, Canada (Vol. 29).

[39] Kurzynowski, T., Pawlak, A., & Smolina, I. (2020). The potential of SLM technology for processing magnesium alloys in aerospace industry. *Archives of Civil and Mechanical Engineering*, 20(1), 23.

[40] Jamróz, W., Szafraniec, J., Kurek, M., & Jachowicz, R. (2018). 3D printing in pharmaceutical and medical applications-recent achievements and challenges. *Pharmaceutical Research*, 35, 1–22.

[41] Dodziuk H., (2020), 'Perspektywy rozwoju druku 3D', Mechanik 01/2020, access on 24.04.2020.

[42] Niemczyk, J., Trzaska, R., Borowski, K., & Karolczak, P. (2019). Scalability 4.0 as economic rent in industry 4.0. *Transformations in Business & Economics*, 18, 69–84.

Chapter 12

Digital twin-driven additive manufacturing

Advancements and future prospects

P M Abhilash
University of Strathclyde

Jibin Boban
Mikrotools Pte Ltd.

Afzaal Ahmed
Indian Institute of Technology, Palakkad

Xichun Luo
University of Strathclyde

12.1 INTRODUCTION

Additive manufacturing (AM) involves layer-by-layer manufacturing of a 3D object from the CAD STL file. AM has drastically changed the manufacturing landscape through its design flexibility and near-net-shape production capabilities. AM makes it possible to conceive new designs and functionalities that were previously unattainable. This has been reflected in its accelerated market growth by several folds in the past decade, and the forecast for the near future is even better. There are several metal AM processes to choose from based on the energy source and metal deposition mechanism, as shown in Figure 12.1. They exhibit a range of capabilities in terms of productivity, part quality compliance, surface integrity, mechanical behavior and chemical properties. However, despite having all these merits and flexibilities over other conventional manufacturing techniques, the inherent complex thermo-mechanical manufacturing mechanism has fundamentally limited the stability of the AM process. Due to its inherent variability and uncertainties, the process has under-par repeatability [1,2]. In addition to the process instabilities, other limitations, such as unanticipated process failures, part defects, poor efficiency and higher production costs, have restricted the widespread adoption of AM components in the aerospace, defense, automobile and biomedical industries, where the accuracy and repeatability of components are vital.

DOI: 10.1201/9781003406488-12

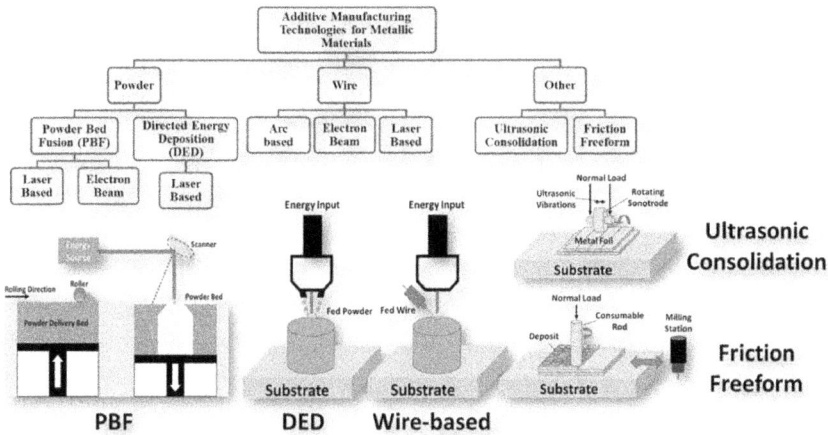

Figure 12.1 Schematic and classification of various metal AM processes [3].

A digital twin (DT) is a virtual replica of a physical entity, process, or system. In the manufacturing context, DTs are used for real-time simulation, analysis and process control toward better performance. DTs have been built to simulate just the product, the production process, or an entire factory environment. This is beneficial since the manufacturers are now able to analyze and optimize their production systems in a virtual environment, reducing the need for costly physical testing and experimentation. Such capabilities of DT can be utilized toward addressing the process uncertainties of AM. Through real-time data acquisition and adaptive process control, DT ensures process certainty by keeping the process within optimal limits in real-time. Also, the DTs can replace conventional feedback approaches with AI-driven feed-forward adaptive process control. This means the future state of the physical AM system is anticipated with respect to the current state, which allows the DTs to proactively perform the precautionary actions as opposed to the conventional reactive strategy of feedback systems. A DT concept aimed at addressing AM process uncertainties is given in Figure 12.2. The process is monitored through multiple sensors, and the type and extent of preventive actions are determined by AI models. The acquired data is used for process diagnosis and real-time updating of the virtual model. Apart from anomaly detection and feed-forward control, DT-driven product design has also gained a lot of research interest in recent years. This concept of Design DT for AM emphasizes a 2-way data communication between the digital design and physical space, to achieve tailored product features including porosity, material properties, geometry and dimensions.

The overall capabilities of DTs toward addressing the challenges of AM are as follows:

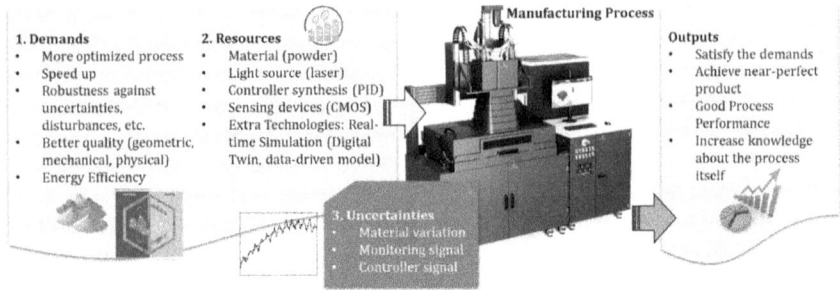

Figure 12.2 A digital twin-driven AM system to address inherent uncertainties [4].

- Process optimization: DT optimizes the AM build parameters for better quality and efficiency.
- Predictive maintenance: DT monitors the AM process performance and forecasts the required maintenance, reducing the overall downtime and enhancing equipment lifespan.
- Quality control: DT simulates the AM build process and identifies potential issues beforehand, ensuring quality control and waste reduction.
- Remote monitoring: DT can remotely monitor the AM process, providing data-driven real-time updates of the build progress and performance.
- Cost optimization: Through DT simulation, the optimal design and process parameters are identified to minimize cost without compromising product quality compliance.

Overall, the DT can be used to anticipate and optimize how a 3D printed product or AM process will behave, find future problems or failures and boost overall effectiveness and efficacy. Upon adopting the DT technology, high quality-low cost AM products can be manufactured with a shorter lead time. In this chapter, the fundamentals, architecture, standards, computing aspects and enabling technologies of AM DTs are critically examined, along with their present challenges, future prospects and research directions. In addition, the chapter further emphasizes enhancing the acceptability of AM across a broader scope of industrial applications due to enhanced robustness and efficiency by employing a DT. Although the DT-driven framework is relevant for all types of AM processes, due to its wider research and industrial significance, DT for metal AM is more stressed in this book chapter.

12.2 COMPONENTS OF DT-DRIVEN AM

A typical DT system comprises (a) physical components, (b) a virtual environment/cyberspace and (c) a communication architecture that connects physical and cyberspace.

- Physical components include the 3D printer and all other associated hardware for data acquisition, communication, computation and storage.
- Visualization models [2D/3D CAD model, augmented reality and virtual reality (AR/VR) visualization], predictive models, simulation models and process control algorithms constitute virtual space. It can also contain auxiliary models for data (signal/image) processing, denoising, feature selection, etc.
- Finally, the communication standards govern and manage the data flow between cyberspace and physical space. The appropriate selection of communication standards and computing architecture is crucial to developing an efficient DT for AM, given the complexity and volume of the data generated during the process.

The overall logical representation of a DT-driven AM process is given in Figure 12.3. The three components are described further in the following subsections.

12.2.1 Physical space

The main element of the physical space is the metal 3D printer itself. The raw materials (metal powder), in situ sensors, communication devices, workstations and equipment for metrology and post-processing are the other constituents of this domain. The right selection of hardware for data acquisition and processing is critical since the accuracy of the data-driven DT depends on the type, number and placement of sensors. The selection of sensor type depends on the type of process signatures to be

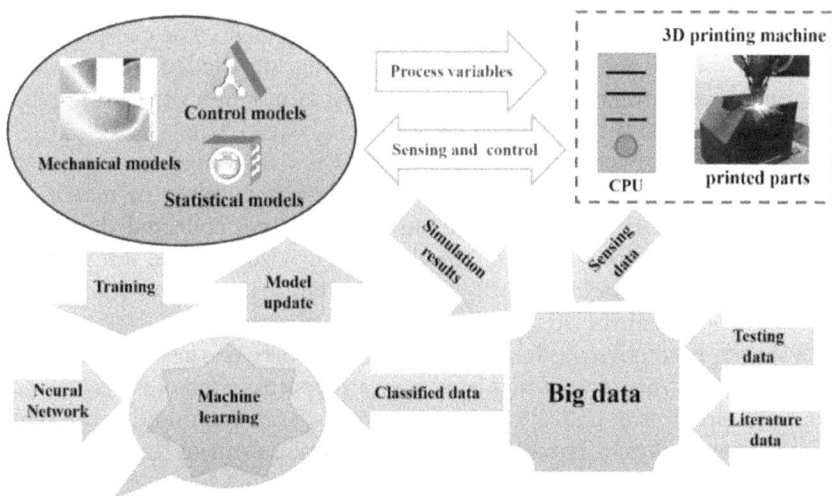

Figure 12.3 Logical representation of DT-driven metal AM [5].

acquired. Sensors can be direct or indirect sensors, depending on their sensing capabilities. The direct sensor captures the feature to be analyzed by acquiring the feature directly. An example of direct sensing is melt pool anomaly detection by imaging the melt pool using a high-speed camera [6]. On the other hand, the indirect sensing method monitors a feature by acquiring secondary signatures. For instance, using acoustic signals to monitor melt pool anomalies. Although indirect sensing is usually easier to set up and analyze, the accuracy can be lower than that of direct sensing methods [7].

A variety of vision-based sensors are used for direct sensing, which includes photodiodes, infrared (IR) cameras, charge-coupled device (CCD) cameras and complementary metal oxide semiconductor (CMOS) cameras. Photodiode and IR cameras capture thermal signals (IR rays) from the in-process melt pool, whereas the CCD and CMOS cameras capture the optical signals in the visible light emission and ultraviolet (UV) emissions. Among indirect sensing, the acquisition of audio and electrical signals is more prominent. The audio signals are captured by acoustic sensors, and additional electrical signals such as eddy currents and instantaneous displacements are captured by the current sensor and strain gauges, respectively [8].

12.2.2 Virtual space

Virtual space includes the data sources, acquisition methods, AM software, AM database, digital twin human-machine interface (DT-HMI) and knowledge base, together with every other digital model, method and algorithm required to run the DT. The domain performs key tasks such as knowledge and data management, process planning, design assistance, tool path generation, online monitoring, diagnosis and prognosis, response prediction, optimization, process control and security management. The computations are based on the available design data, existing knowledge database and real-time acquired data. Some of the critical data sources within a metal AM process are given in Figure 12.4.

Another key aspect of the virtual domain is data visualization or representation. Typically, 3D CAD models update themselves to match the physical state of the system based on the real-time collected data. The other prominent modes of data simulation and visualization are given in Figure 12.5, which include multi-physics simulations, online or offline flaw representation, real-time data monitoring and information analytics. In DT, data-driven predictive models are employed to forecast the future state of the AM process based on which feed-forward process control is performed. Common predictive models are AI/ML models, statistical models, analytical equations, physics-based expressions, or finite element analysis (FEA) simulations. In case where anomalies or defects are predicted, process control models will autonomously perform parameter revisions to regulate and restore AM process stability.

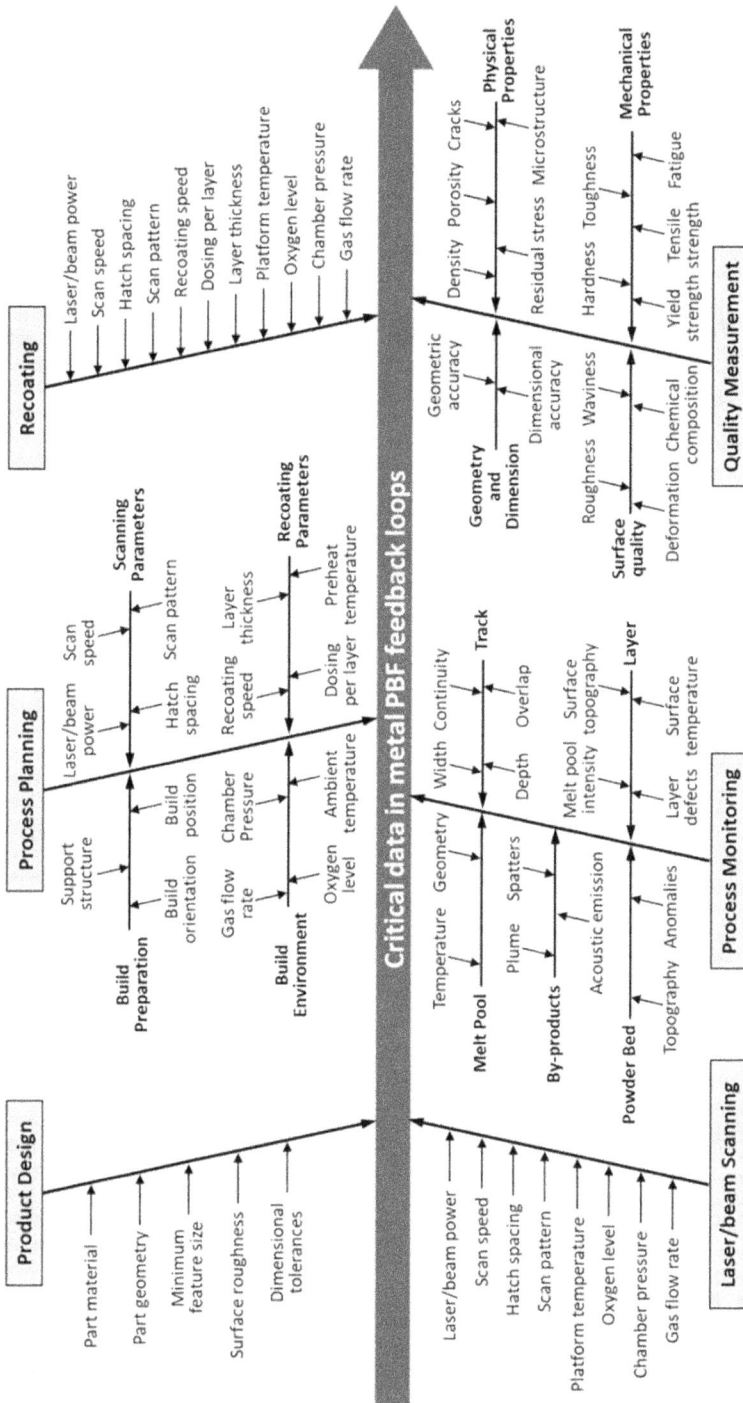

Figure 12.4 Critical data sources at various lifecycle stages in the metal AM process [9].

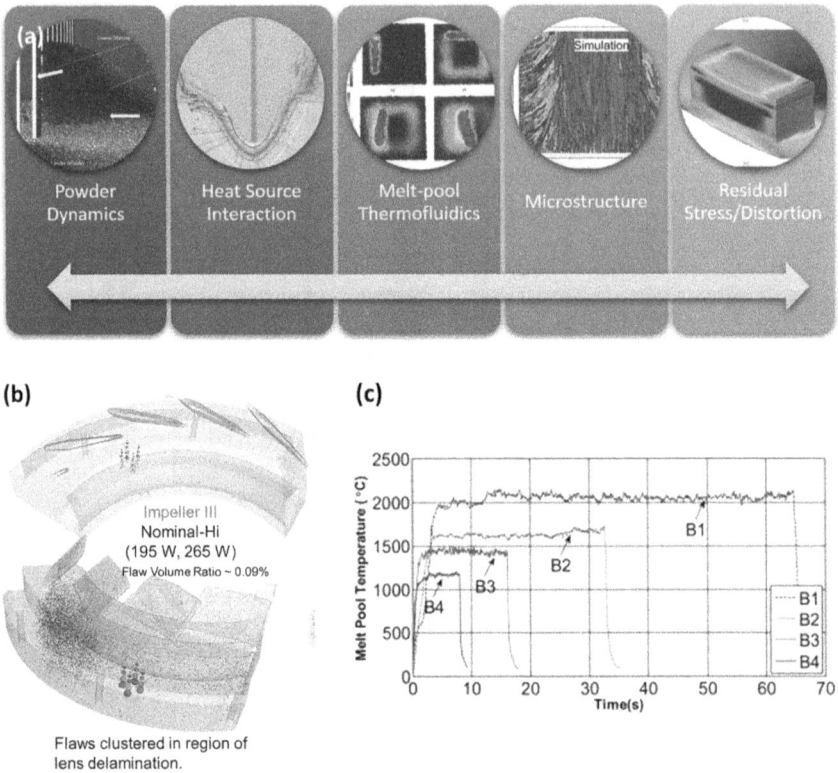

Figure 12.5 (a) Multi-physics simulations [11]. (b) Defect/Flaw visualization [12]. (c) Real-time sensor data [13].

In most DTs, a highly interactive human-machine interface (HMI) is preferred for effective visualization and appropriate human intervention. In this aspect, AR/VR have gained massive public, industrial and research attention in the past decade. It is an advanced digital environment that permits human interactions with virtual objects. In metal AM, such a virtual environment that offers interactive training, moderated by an expert through cloud computing has already been developed [10]. Now that the proof of concept is established, AR/VR have massive potential to leverage the possibilities of over-the-air (OTA) training, remote support and collaborative digital work environments.

12.2.3 Communication standards and data management system

Lack of standardization is one of the key bottlenecks that restricts the cross-compatibility of DT systems. To address this issue, the ISO 23247 standard was developed, which defines the general guidelines for developing a DT framework for manufacturing applications. Based on this standard, a generic DT implementation framework for wire arc AM (WAAM) was developed by Kim et al. [14]. The proposed framework is aimed at the customizability and interoperability of DT. To demonstrate the framework, a case study of online anomaly detection and process control has been considered. Data collection and communication are well defined, with different moduli for data collection, device control computing and HMI. The core entity for data monitoring and analysis is proposed through edge, cloud, or fog computing technologies. Cloud computing promotes centralized data handling but is less secure, demands greater data handling and can often result in prediction latency. On the other hand, edge computing performs distributed data analysis and is more secure and fast. Fog computing is a hybrid technology with a combined advantage over the other two.

Another collaborative data management system for DT-driven AM is shown in Figure 12.6 [15]. Each module uses edge computing to perform specific tasks and also communicates with the cloud storage for centralized storage, retrieval and collaborative data communication. The data generated in an AM process lifecycle comes mainly from five sources: design

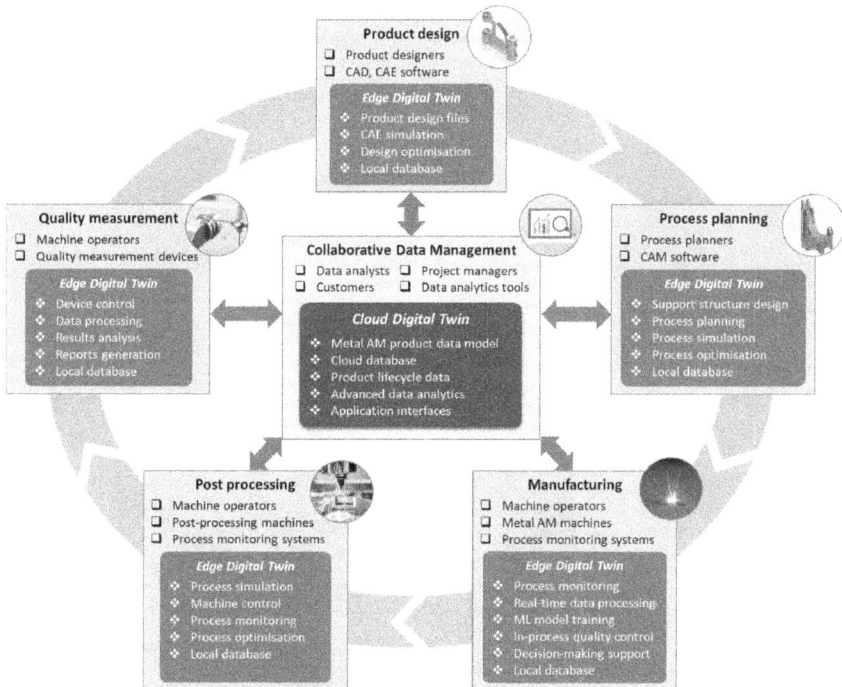

Figure 12.6 Concept of a data management and communication framework for DT-driven AM [15].

data, process parameters, online monitoring data, post-processing data and metrological data. During the design stage, the required part is designed as a CAD file, and appropriate offline simulations are performed. The design digital model is shared with the cloud to be retrieved as and when required by the other modules. For example, the design data may be accessed to fix the scan strategy during post-processing and final product quality evaluation. Ren et al. [16] have also presented an edge-cloud data flow framework for a metal AM system.

12.3 ROLE OF AI AND COMPUTING STRUCTURE IN DT-DRIVEN AM

Artificial intelligence (AI) has a key role to play in building the DT of AM processes. AI is extensively utilized in three aspects of AM, namely, design, process and production. Figure 12.7 represents the AI-driven tasks during each of these stages. Even in the past, AI models have been widely used to enhance process efficiency and the overall performance of complex manufacturing processes. However, thanks to several recent technological advancements in AI, there has been exponential growth in its computing power, efficiency and data-handling potential. Thus, AI-driven DTs can now accurately and efficiently simulate even complex processes like AM

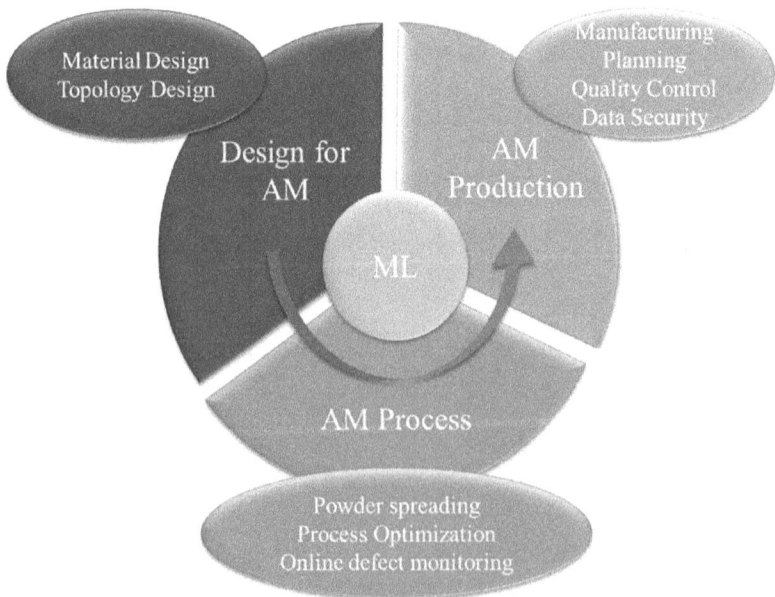

Figure 12.7 Application areas for ML-based DT in metal additive manufacturing [17].

in real time. Apart from response prediction, AI models perform online sensor-data analytics for anomaly/defect/error detection, prediction, classification and rectification. Apart from AI technologies, appropriate selection and usage of various computing architectures such as edge, fog and cloud computing have a critical influence on DT performance in terms of data latency, security and storage requirements. In short, advancements in AI and computing structures have tremendously influenced the capabilities of DT AM processes toward better productivity, quality and cost-effectiveness for manufacturers. Some of the prominent AI learning schemes and computing architectures are described in the following subsections in the context of DT-driven AM.

12.3.1 Supervised ML for data-driven real-time predictions

Supervised learning is an ML approach that trains a model based on a labeled dataset. For an AM training dataset, the process parameters are the model inputs, and the output is the desired response (like part quality, or productivity). Once trained, the model will make predictions on unseen data. When such models are used within a DT, the AM outcomes are predicted in real time, which allows for subsequent process adjustments if required. Such predictive models improve process efficiency by minimizing the number of trial runs. A critical point to note here is that, for a supervised ML model, the prediction accuracy will be very much reliant on the quality of the training data. It is thus important to acquire a diverse range of training data to ascertain the robustness of the model, which may otherwise lead to poor predictions. Some of the widely used supervised ML models are convolutional neural networks (CNN), support vector machines (SVM) and artificial neural networks (ANN). These models are briefly discussed below:

- ANN models are well suited to model AM processes due to their proven ability to handle non-linear data, large datasets, missing data and noisy data. Moreover, it is a robust algorithm well suited for real-time predictions such as anomaly detection and control systems [18,19]. Some of the limitations are the presence of a large number of hyperparameters to train and select, a longer training duration and poor explainability.
- CNNs are a type of ANN that is particularly well suited for image and video recognition tasks. These models can be used in AM DTs for quality control and inspection, predictive maintenance, process monitoring, simulation and virtual prototyping and automated surface reconstruction. In addition, other types of neural networks can be used in combination with CNNs to improve the performance of the model [20].

- SVM is considered more robust against overfitting than neural networks, is more adaptable and has superior generalization capabilities. Due to an interpretable decision surface, it can be regarded as more explainable as well. Also, due to its computational efficiency, the model is well suited to handle complex and data-rich processes such as AM [21].

12.3.2 Cloud/Fog/Edge computing

Computing power, data security, data storage and data access can be put to good use when DTs are run on cloud, fog, or edge computing platforms. Cloud computing refers to a computing technique in which the entire system's data is stored in a centralized online storage space (cloud) and retrieved from the cloud on demand. Although there are several advantages, such as better resource sharing and collaboration leading to reduced setup expenses, cloud computing is not ideally recommended for real-time systems due to data latency, poor security and difficulties in remote access. In contrast to this approach, edge computing architecture decentralizes the computations by taking them nearer to the data source or the user, resulting in quicker responses. Apart from reduced latency, edge computing offers enhanced security and privacy, reduced operational expenses and better reliability. Fog computing is a recent hybrid concept designed to address the limitations of cloud computing by moving the cloud toward the edge of the IoT network. Appropriate integration of these computing techniques is vital to enhancing the computing capabilities of a DT platform. The usage of these technologies in DT-driven AM is presented and discussed in this section (Figure 12.8).

Integration of cloud and edge computing can be demonstrated via a multi-level DT architecture for metal AM [16]. Various levels in this hierarchy are the machine level, centralized factory (cloud level) and research center (edge level). Optimal feature selection through dimensionality reduction is extremely vital in an edge-cloud data communication scheme. This allows the elimination of data redundancy and reduces the latency of the DT system. Part design data (called recipes) is communicated to the machine from the factory cloud. During 3D printing, process signatures acquired by sensors are analyzed to check for build errors and process

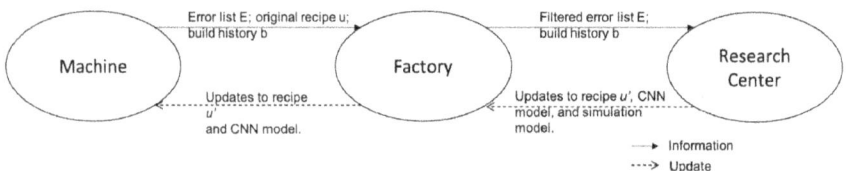

Figure 12.8 Edge-cloud data communication guidelines for metal AM systems [16].

anomalies. The errors are listed and validated, and key findings are communicated back to the cloud and research levels. Raw data and error lists are down-sampled during their transfer to minimize latency, storage and bandwidth requirements. Also, the error list communication to the cloud is performed only upon encountering a new anomaly to expand the database. This information is incorporated into the subsequent part design. On repeated occurrences of process anomalies, it is sent to the research level for detained analysis and further decision-making, including a processing halt. The overall data-transfer scheme is shown in Figure 12.8. A similar multi-layer computing scheme was proposed by Guo et al. [22], but for the fused deposition modeling (FDM) process. The device, edge and cloud layers are responsible for data acquisition, local data processing and big data analysis, respectively. Recently, another noteworthy cloud-edge computing scheme for DT-driven AM was proposed by Liu et al. [15]. This data management system comprises a centralized cloud DT and multiple-edge DTs, one per lifecycle module. Each lifecycle module and distributed shop floors communicate the data to the cloud DT, where data analytics is performed. A dedicated model handles all the data associated with product quality compliance at different lifecycle stages. The collaborative cloud-edge DT data management system enables offline CAD+STL design, layer-wise defect detection and parameter optimization.

A comprehensive IoT-driven cloud computing platform integrating design, 3D printing and production planning is presented by Wang et al. [23]. Cloud computing integrates and manages information flow across hardware and software modules. This not only includes the data from 3D printers, materials, sensors and software but also expert knowledge and technical know-how. The paradigm offers a wider scope for end users and stakeholders to monitor and control the process remotely. IoT architecture facilitates information flow across design, processing and planning, thereby reducing product development bottlenecks. Adnan et al. [24] proposed fog computing-based real-time closed-loop process control for AM processes. The proposed computing architecture is a hybrid mix of cloud, edge and fog technologies. The fog computing structure addresses the fundamental limitations of the cloud-based platform in terms of latency, bandwidth requirements, seamless integration with cloud services, data communication and security.

12.3.3 Unsupervised and reinforcement learning

One of the major drawbacks of supervised learning models is the requirement for a training dataset. Often, the AM process is stochastic and can cause new anomalies outside the ones recorded during the training in real-world test situations. To overcome this fundamental limitation of supervised ML models, unsupervised and reinforcement learning schemes are being used. One of the key applications of unsupervised learning is defect detection. In this regard, an effective strategy is to record the process signatures during both defective and normal operations and then utilize a self-organizing map (SOM)-driven online monitoring to distinguish between various process anomalies [25]. Any appropriate AM process signatures can be acquired and processed in this regard, including thermal signatures, acoustic emission, or vibration data. Although clustering is a very capable method to group process signatures/process histories based on process signatures, the approach may be unsuitable to operate within an AM DT, where it can suffer from prediction delays while handling the raw data from the sensors. A more computationally efficient and robust approach to unsupervised anomaly detection is to perform dimensionality reduction, followed by clustering. The approach is generic and can be incorporated with any appropriate selection of process signatures, dimensionality reduction method and clustering algorithm based on the type of AM process and the end application. Dimensionality reduction reduces the order of a multi-dimensional feature space, and it can significantly reduce the latency issues during a real-time DT operation. This approach has been demonstrated successfully for laser powder bed fusion by acquiring computer vision-based process signatures. A t-SNE (t-distributed stochastic neighbor embedding) dimensionality reduction algorithm is used to convert the raw multi-dimensional datasets into low-dimensional 2D feature space data. The 2D data is then clustered using K-means to visualize, detect and identify process flaws such as porosity, undermelting and balling [6]. For unsupervised clustering, K-means clustering is one of the simplest yet robust techniques to group the feature sets into k groups based on geometric distance [26]. Later, a similar approach was chosen to analyze time series thermal signatures. Here thermal processing history zones are mapped into a predefined number of clusters using a K-means clustering algorithm, following dimensionality reduction by symbolic aggregate approximation (SAX). The thermal zone mapping contributes to the better interpretability of material processing, considering various scan paths and geometries [27]. The unsupervised clustering algorithm is utilized for several other applications within metal AM. It is used to categorize wire bead consistency in WAAM [28]. Recently, K-means clustering has been used for surface quality categorization during direct energy deposition (DED) AM, in which surfaces are classified based on porosity, surface morphology and deposition quality by acquiring the emission spectrum [29].

12.4 KEY APPLICATION AREAS OF DT IN AM

12.4.1 Process condition monitoring and quality control

The integration of ML with DTs in real-time aids in the automatic and rapid analysis of massive data with scalable performance. Reliable predictions of process states and defects can be accomplished through ML models that ultimately pave the way for effective process monitoring of the AM process. The DT framework proposed by Mourtzis et al. [30] for the FDM process is noteworthy for monitoring the process and optimizing the parameters. The DT architecture in the study is formulated to enhance AM product quality and minimize cost in the long run. Machine operation and process monitoring can be realized through the developed AR-based immersive GUIs. Later on, the engineer executes a quality assessment and stores the results for future reference in a cloud database, which serves as an efficient resource for process parameter optimization. Based on the process deviations observed during online monitoring, the process parameters can be adequately controlled to maintain the desired part quality.

12.4.2 Detection of process defects/anomalies

In the current industrial scenario, the AM domain is greatly dependent on machine learning (ML)-assisted DTs for the detection of flaws or defects in fabricated components. Despite traditional image processing methods, in situ cloud processing techniques using ML yield viable results in the identification of AM component defects or anomalies [31,32]. A software platform capable of executing multiple sub-processes in parallel enables the in situ cloud processing of data. The ML model incorporated in the cloud processing node facilitates the real-time identification of surface defects and anomalies with high accuracy. Thus, DT approaches can be adopted for online monitoring and flaw detection in the LPBF process. Yavari et al. [12] realized successful flaw detection in LPBF SS316L impeller-shaped components through the DT approach. The in situ measurements of melt pool temperature are combined with a thermal computational model capable of predicting the temperature distribution in the component. The predictions of the model were updated simultaneously, layer by layer, with the in situ measurements of melt pool data. The proposed approach proved successful in detecting process parameter changes/drifts, implanted voids and delamination, thereby contributing to the precise detection of flaw evolution.

12.4.3 Output prediction of AM process attributes

Enormous experiments have to be conducted to study the behavior of the AM process, which involves time-bound effort. Such experimental studies involve the wastage of samples in large quantities due to trial-and-error approaches and non-destructive testing procedures [5]. DT technology facilitates the accurate prediction of output response in the AM process, comprising microstructural aspects, mechanical properties, surface characteristics and dimensional deviations. DT makes use of the thermal, structural and material models for predicting the output responses. In light of such models, DT provides detailed information on the evolution of defects, the development of microstructure, temperature gradients existing over build volume and the optimal combination of process parameters for the best part quality. Industries demanding AM parts/components strongly rely on DT-supported prediction models for enhanced product quality. In addition, the DT approach ensures faster prediction with minimal time and without material waste.

12.4.4 Tool path planning and simulation

The manufacturing domain has advanced to the point where production layout information can be communicated with the assistance of AR between a reconfigurable AM system composed of robotic arms and the associated DT for tool path generation [33]. The laser scan strategies in metal AM can be optimized to fulfill faster production rates and prevent instabilities in fabrication. Spatial relations among the various system attributes comprising a part substrate, robotic arms, camera, etc. are obtained in the form of transformation matrices to provide information regarding production layout. The same layout information is provided as input into the DT simulation, which ensures the convenient implementation of an optimized layout in the physical system. The deployment of an optimized layout can shorten the processing time and minimize the costs associated with production.

12.4.5 Data management for manufacturing systems

From the perspective of Industry 4.0, DT can be considered a promising technology for smart manufacturing systems and cyber-physical systems (CPSs). The islands of information existing between various stages of the product life cycle (due to different data types) as well as the lack of iteration/ interaction between different activities in the lifecycle can hinder proper data management in production systems [34]. However, recent innovations in the DT approach exhibit immense potential for supporting efficient data management by overcoming the aforementioned challenges. Developing DT in the cloud can open up an effective environment for data management that supports various types of advanced data analytics, such as the real-time status of the shop floor, machine condition monitoring and production information [15]. Moreover, DT functions as a powerful platform for the exchange of data/information among sensors and controllers, thereby facilitating fruitful data visualization and analysis. Thus, the interoperability of data can be successfully realized in AM systems using a DT-driven approach.

12.4.6 Post-processing operations

DT technology is also adopted in the post-processing operations of AM components. Post-treatment operations are inevitable after AM fabrication to improve the part quality in terms of surface integrity, form accuracy and mechanical properties [35,36]. The post-treatment operations can be effectively automated by utilizing the AM dataset contributed by DT, which comprises process parameter data, fabrication methodology and material design. Thus, the automation efforts in post-treatment phases can be reduced with the assistance of AM DT. Moreover, enhanced quality of AM parts/components can be assured by integrating DTs of respective post-processing operations with AM DTs.

12.4.7 Digital visualization of physical processes

DT enables the visualization of the AM process via the complete modeling of the process in a virtual world. The physical process and virtual model are linked to work together simultaneously, which enables the proper control of the process through a digital platform. The virtual demonstration using 3D CAD model-based simulations favors a better understanding of the state and dynamics of the actual AM process. Thus, process visualization using DTs allows humans to interact with the model, analyze it and suggest appropriate changes, thereby saving cost and time associated with production. Moreover, industries can assess optimizations at desired phases in the process cycle by watching DT-enabled process visualization on screens.

12.4.8 Design optimization of product

DT facilitates the creation of a comprehensive and integrated data model that continuously updates the attributes of the live process in real time, thereby opening up the opportunity for effective product design. High-performance IoT sensors, data storage and transmission systems enable the collection of product operating data and provide feedback into the digital product model, followed by closed-loop control of the real-time process. The traditional way of designing products is fundamentally based on the data obtained from simulation using historical or estimated values. Such a procedure often leads to uncertainty in product design. Consequently, the engineers overcome the risk in design by incorporating high safety factors, resulting in ineffective resource utilization and the development of products with larger dimensions. The investigation conducted by Bellalouna et al. [37] in evaluating the efficiency of DT in optimizing product design is noteworthy in this regard. DT technology is employed for obtaining load and material stress data accurately from product operation in order to use the data for efficient and load-oriented product design. In the study, the FEA of the product is executed based on operating data rather than any historical or estimated data. Thus, dependable load analysis using operating data aided product topology optimization and effective material utilization. The key application areas of DT in the AM domain are portrayed in Figure 12.9.

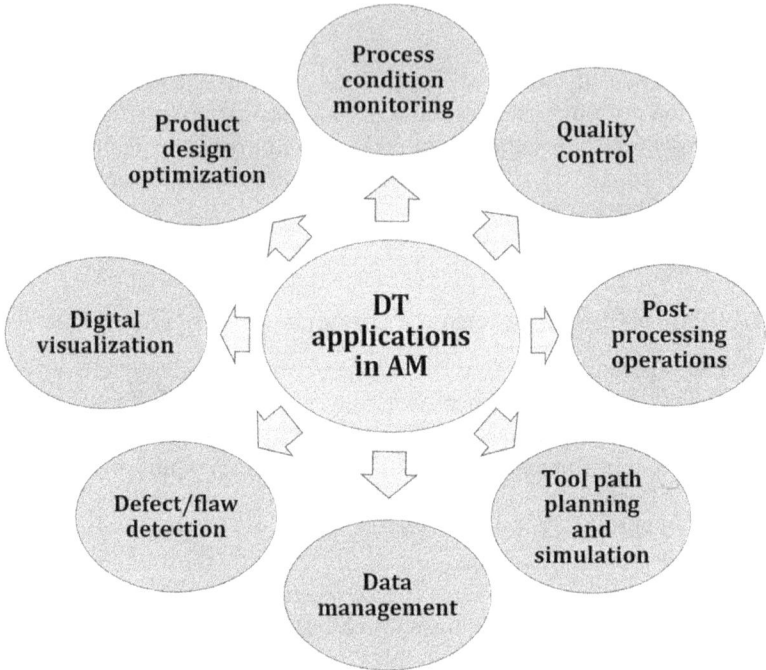

Figure 12.9 Application areas of DT technology in AM.

12.5 CHALLENGES AND FUTURE SCOPES

Challenges: Although DT-enabled AM process control can bring promising outcomes in the industrial domain, several challenges hinder the successful deployment of DT technology in practical applications. Such challenging factors are limited explainability, lack of standardization in DT structure and data communication, privacy and security vulnerabilities, computing power requirements, prediction latency, etc. that need to be critically addressed for proposing suitable solutions.

- *System complexity and uncertainty*: The DT integration on AM is challenging due to the complexities and uncertainties associated with the process. For a metal AM process, knowledge of uncertainty sources is very limited so far, which in turn affects its effective modeling as well. Overall DT system uncertainty will exponentially shoot up when this process-level uncertainty gets superimposed on the intrinsic computational uncertainties [38].
- *Lack of standardization*: Integration of the CPS is another critical bottleneck. The hardware and software of various components in CPS are all heterogeneous in nature. It is a huge task for them to

be integrated via standard physical and communication structures so that cross-domain synergy is achieved throughout the product lifecycle.

Apart from the lack of standardization within a DT, another biggest challenge is the lack of inter-DT cross-compatibility. Each DT is different from others in its architecture and functioning due to the lack of standardization. Standardization facilities smoother data communication, hardware cross compatibilities, enhanced safety and structural integrity. The time and cost to develop DT are not justifiable for AM if it is compatible with a single system. To tackle this issue and widen the scope of DT implementation for AM, multiple organizations, such as ISO and IIC, are working toward a standard protocol for DTs. But this has opened up a new challenge since several independent organizations are now working toward the same problem without formal alignment or collaboration. Thus, even after these standards are implemented, they may be very different from one another, leaving the problem of interoperability unaddressed or partially addressed [39].

- *Data Challenges*: Due to advancements in sensing technologies, a huge volume of data will be acquired by real-time DTs every second. This is especially true for an AM process since vision-based monitoring is one of the most preferred sensing techniques. Not only does real-time image processing demand huge computational power, but extracting useful information from such large and complex data is also difficult. In AM, most of the physical quantities of interest, such as melt pool temperature, heat distribution, and stress distribution, are computationally intensive to simulate and predict. The cumulative effect of computational complexities, ineffective data acquisition, storage, analytics and communication is significant prediction latencies leading to inaccurate conclusions and delayed process controls [5]. In the realm of AR/VR, the act of reducing the size of extensive datasets to enhance computational efficiency frequently results in a diminished quality of the rendered 3D experience. Selection of the most appropriate computing structure (edge, cloud, or fog) and dimensionality reduction can reduce the effect of big data. In any case, real-time data transmission through wireless mode is challenging due to the restricted sampling rate, bandwidth and storage technology [40].

In addition, there are a lot of concerns about the privacy, security and data-sharing aspects of DT. Since DT is an online, connected and collaborative environment, there are plenty of privacy and security threats. The major ones are unauthorized system access, IP hacking and non-standard security compliance. Cross-device cloud-based information sharing creates legal vagueness about data ownership. The critical data-related challenges associated with DTs are consolidated in Figure 12.10.

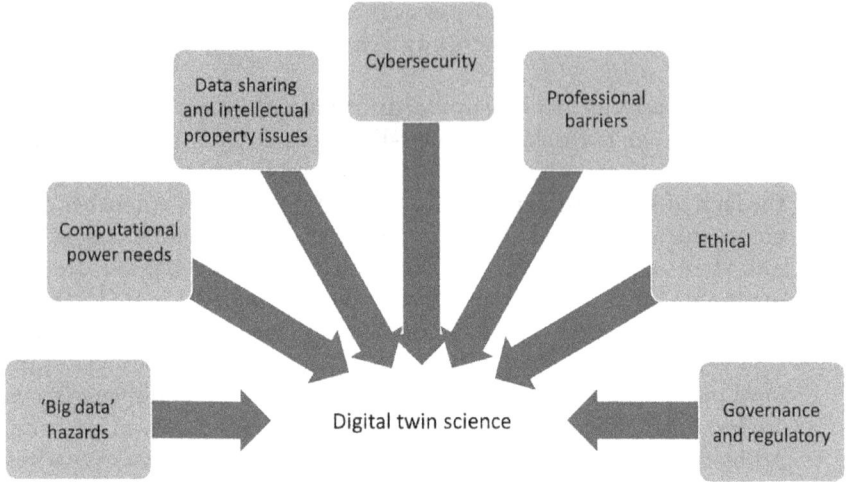

Figure 12.10 DT data-related challenges [41].

- *Explainability*: The AI-ML models within a DT consider multi-format data (images, signals, videos and 2D/3D renderings) for various tasks such as anomaly prognosis, diagnosis, response prediction, process control, tool path generation, or in-process optimization during the AM process. The complicated computations involved in these predictions have demanded the use of advanced and sophisticated AI models. However, since these models offer a black-box-style prediction with no feedback on the prediction logic, it is challenging to convince the stakeholders to trust such a DT system [42]. Also, due to their opaqueness, the current DT-driven AM systems are targeted toward highly proficient domain experts, with regular operators with semi-technical knowledge finding it challenging to take advantage of the DT capabilities. Although a newer field of study is being introduced to impart interpretability, called explainable AI (XAI), its application scope is currently limited to very few manufacturing problems. A typical XAI system for the AM defect prediction model is given in Figure 12.11. That being said, the seamless integration of XAI tools with complex and multi-model systems like DT is very challenging and yet to be implemented.
- *Scalability*: The scale of DT has several contradictory points of view. The scope of DT can be as simple as an offline system with intermittent data communication. However, many argue that 2-way real-time communication is a must for a system to be considered DT. Again, there are contradictions on whether an online CAD visualization is mandatory. Also, DT can be built into a simple process (online monitoring of the melt pool during the AM process) or product (inline

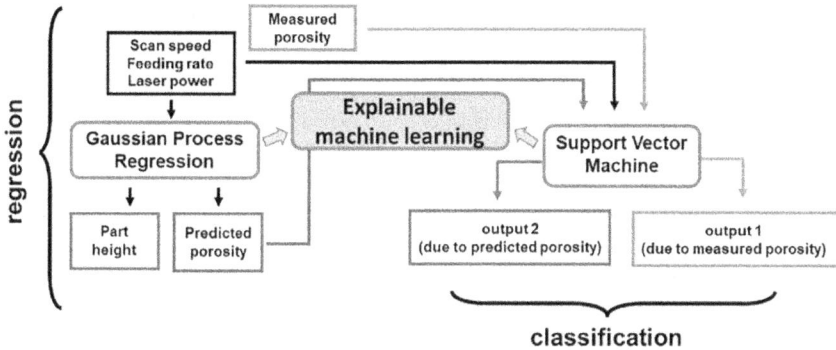

Figure 12.11 An explainable AI (XAI) architecture for a metal AM anomaly detection and prediction system [43].

part anomaly detection) and can extend up to a highly complex smart factory (starting from design all the way through metrology and inspection). These ambiguities regarding the scope and scale of DTs have often resulted in the development of non-robust DT systems. Such DTs are not scalable to a smaller or larger context. The generalization and robustness of the system-level DT architecture need further improvement to achieve varied applications, viewpoints and contexts [44].

• *Poor integration of human factors*: Although completely autonomous systems capable of zero-defect manufacturing are the vision of Industry 4.0, the present state of technology requires human involvement at many critical junctures. However, unlike digital systems, human operators cannot be modeled or monitored. This results in DTs with either zero space for human intervention or DTs that get halted until a human operator gives his inputs. A human DT is a compromised solution to these extremities [45], but modeling human-specific features such as fatigue, mood and physical attributes is extremely challenging in an AM environment.

Future scope: In the present manufacturing context, the DT-driven AM approach seeks significant advancement in many areas to ensure wide acceptability and applicability in industrial applications. The potential topics of fruitful research in the future consist of uncertainty reduction in DTs, novel data management strategies, explainable AI-driven DTs, robustness enhancement, etc.

• *Elimination of uncertainty*: The AM process involves extremely complex metallurgical changes and thermal physics. Thus, the representation of the AM process using DT models often accepts uncertain input data owing to the challenges in accounting for all the process parameters and physical phenomena. Therefore, a more mature data-driven framework with timely calibrations and seamless integration with multiple lifecycles is required to minimize the uncertainty of the AM DT.

- *Narrowing the interoperability gaps*: Immense scope persists in reducing the interoperability gaps in DT by standardizing the DT architecture. Extensive adoption of existing standards can favor interoperability in the DT system. Therefore, it is meaningful to compare the structural framework of various DT standards and evaluate the corresponding alignment and contradictions for successfully overcoming interoperability issues.

- *Strategies for data management*: Future research thrust can be centered on better data-handling strategies aimed at faster data transfer without compromising on data quality or losing out on information. Similarly, research focus can be provided to develop better data frameworks, architectures and computing models for effective data utilization. Superior computationally efficient models and dimensionality reduction techniques are in huge demand for creating a structured and systematic data management system. Moreover, more effective tools for data analysis and management will make DT models more scalable in a large spectrum of manufacturing applications.

- *Process and product FP identification*: Process and product FP identification is an emerging and futuristic research area with a core emphasis on extracting the fundamental relationships between process parameters and their functional performance. Developing such models with the support of DT looks promising, as it enables computationally efficient monitoring and quality control during the AM process. Process FP-driven DT and real-time product FP control can transform the way DTs are perceived until this point.

- *ML and AI*: ML and AI algorithms are capable of greatly enhancing the effectiveness of DTs in the optimization and prediction of the AM process. Research concentration can be directed toward incorporating ML and AI into DT models to transform them into more self-learning and intelligent models.

- *Explainability of systems*: Research direction toward improving the explainability of DT-driven AM systems exhibits vast potential ahead. DT predictions and actions for an AM process need to be explained in a human-comprehensible manner to bring trust and transparency to the system. In the future, deep integration of XAI with every DT predictive model should be incorporated in such a way that the decision-making rationale at every step is made clear to the expert/stakeholder. A high degree of system explainability assists the expert in decision-making and implementing required system/process improvement changes.

- *Data privacy and security*: A massive scope of research interest is reflected in the areas of DT data privacy and security. Risk assessment studies for DT constitute one of the first steps in this regard. These include asking questions about the kind of data sharing and communication necessary for the application. Data can be classified into in-built (from AM systems and default controllers) and acquired (from IoT sensors), followed by appropriate data governance, allocating variable data access and security. For instance, as several IPs may not require central cloud access, IP theft can be minimized by restricting access, thereby ensuring enhanced data privacy and security.
- *Robustness of DT in AM processes*: The first generation of DT-driven AM continues to be in the development phase in the manufacturing domain. The existing DT models are limited to only a few AM methods, materials and equipment. Hence, significant research focus can be set on improving the robustness of DT to work across various AM machines, materials and processing technologies. Advanced material models have to be developed for AM, which takes into consideration factors such as defects, microstructure, part distortion and design deviations that affect the mechanical properties of the component. Further, the proper integration of AM systems with advanced algorithms and databases can yield promising improvements in the field of AM.

12.6 CONCLUSIONS

Digital twin technology is extremely capable of transforming the AM industry by addressing existing challenges and unraveling its future potential. DT provides the manufacturers with the flexibility to simulate, predict and control the process with minimal trial-and-error or experimental testing. The technology will shape the future of the AM field by offering insightful information on the process through better understanding, representation and analysis of the physical AM process. The integration of AI/ML tools such as ANN, CNN and SVM further enhances the DT capabilities through better data analytics and predictive capabilities. In short, the adaptation of DT will massively improve the process efficiency, productivity, quality and cost-effectiveness of AM for manufacturers and stakeholders. Moreover, the capability of online virtual representation, monitoring and adaptive control will significantly leverage the reliability of the process.

Notwithstanding the fact that developing a DT for AM is still a challenging task that demands a significant volume of data, computational power and communication resources. In addition, the scope and utilization of DT are very much dependent on the available data and end application. In summary, DT in AM has a promising future owing to the rate of advancement in technologies, which will enable it to overcome the current challenges and enhance the overall repeatability, robustness and acceptability of AM.

ACKNOWLEDGMENTS

The authors gratefully acknowledge the financial support from the UK Engineering and Physical Sciences Research Council (EPSRC, EP/ T024844/1, EP/V055208/1, W004860/1).

REFERENCES

[1] Boban, J., Ahmed, A., 2021, Improving the surface integrity and mechanical properties of additive manufactured stainless steel components by wire electrical discharge polishing, *Journal of Materials Processing Technology*, 291:117013, https://doi.org/10.1016/j.jmatprotec.2020.117013.

[2] Boban, J., Ahmed, A., Rahman, M. A., Rahman, M. A., 2020, Wire electrical discharge polishing of additive manufactured metallic components, *Procedia CIRP*, 87:321–326, https://doi.org/10.1016/j.procir.2020.02.023.

[3] Bandyopadhyay, A., Zhang, Y., Bose, S., 2020, Recent developments in metal additive manufacturing, *Current Opinion in Chemical Engineering*, 28:96–104, https://doi.org/10.1016/j.coche.2020.03.001.

[4] Stavropoulos, P., Papacharalampopoulos, A., Michail, C. K., Chryssolouris, G., 2021, Robust additive manufacturing performance through a control oriented digital twin, *Metals*, 11/5:708.

[5] Zhang, L., Chen, X., Zhou, W., Cheng, T., Chen, L., et al., 2020, Digital twins for additive manufacturing: a state-of-the-art review, *Applied Sciences (Switzerland)*, 10/23:1–10, https://doi.org/10.3390/app10238350.

[6] Scime, L., Beuth, J., 2019, Using machine learning to identify in-situ melt pool signatures indicative of flaw formation in a laser powder bed fusion additive manufacturing process, *Additive Manufacturing*, 25:151–165, https://doi.org/10.1016/j.addma.2018.11.010.

[7] Sturm, L. D., Albakri, M. I., Tarazaga, P. A., Williams, C. B., 2019, In situ monitoring of material jetting additive manufacturing process via impedance based measurements, *Additive Manufacturing*, 28:456–463, https://doi.org/10.1016/J.ADDMA.2019.05.022.

[8] Lin, X., Zhu, K., Fuh, J. Y. H., Duan, X., 2022, Metal-based additive manufacturing condition monitoring methods: from measurement to control, *ISA Transactions*, 120:147–166, https://doi.org/10.1016/j.isatra.2021.03.001.

[9] Liu, C., Roux, L. Le, Ji, Z., Kerfriden, P., Lacan, F., et al., 2020, Machine Learning-enabled feedback loops for metal powder bed fusion additive manufacturing, *Procedia Computer Science*, 176:2586–2595, https://doi.org/10.1016/j.procs.2020.09.314.

[10] Mogessie, M., Wolf, S. D. V., Barbosa, M., Jones, N., McLaren, B. M., 2020, Work-in-progress-a generalizable virtual reality training and intelligent tutor for additive manufacturing, *Proceedings of 6th International Conference of the Immersive Learning Research Network*, iLRN 2020, pp. 355–358, DOI:10.23919/ILRN47897.2020.9155119.

[11] Phua, A., Davies, C. H. J., Delaney, G. W., 2022, A digital twin hierarchy for metal additive manufacturing, *Computers in Industry*, 140:103667, DOI:10.1016/j.compind.2022.103667.

[12] Yavari, R., Riensche, A., Tekerek, E., Jacquemetton, L., Halliday, H., et al., 2021, Digitally twinned additive manufacturing: detecting flaws in laser powder bed fusion by combining thermal simulations with in-situ meltpool sensor data, *Materials & Design*, 211:110167, https://doi.org/10.1016/J.MATDES.2021.110167.

[13] Farshidianfar, M. H., Khajepour, A., Gerlich, A. P., 2016, Effect of real-time cooling rate on microstructure in laser additive manufacturing, *Journal of Materials Processing Technology*, 231:468–478, https://doi.org/10.1016/J.JMATPROTEC.2016.01.017.

[14] Bong Kim, D., Shao, G., Jo, G., 2022, A digital twin implementation architecture for wire + arc additive manufacturing based on ISO 23247, *Manufacturing Letters*, 34:1–5, https://doi.org/10.1016/j.mfglet.2022.08.008.

[15] Liu, C., Le Roux, L., Körner, C., Tabaste, O., Lacan, F., et al., 2022, Digital twin-enabled collaborative data management for metal additive manufacturing systems, *Journal of Manufacturing Systems*, 62:857–874, https://doi.org/10.1016/j.jmsy.2020.05.010.

[16] Ren, Y. M., Ding, Y., Zhang, Y., Christofides, P. D., 2021, A three-level hierachical framework for additive manufacturing, *Digital Chemical Engineering*, 1/June:100001, https://doi.org/10.1016/j.dche.2021.100001.

[17] Wang, C., Tan, X. P., Tor, S. B., Lim, C. S., 2020, Machine learning in additive manufacturing: State-of-the-art and perspectives, *Additive Manufacturing*, 36/August:101538, https://doi.org/10.1016/j.addma.2020.101538.

[18] Abhilash, P. M., Chakradhar, D., 2021, Wire EDM failure prediction and process control based on sensor fusion and pulse train analysis, *International Journal of Advanced Manufacturing Technology*, 118/5–6:1453–1467, https://doi.org/10.1007/s00170-021-07974-8.

[19] Abhilash, P. M., Chakradhar, D., 2022, Performance monitoring and failure prediction system for wire electric discharge machining process through multiple sensor signals, *Machining Science and Technology*, 26/2:245–275, https://doi.org/10.1080/10910344.2022.2044856.

[20] Li, X., Jia, X., Yang, Q., Lee, J., 2020, Quality analysis in metal additive manufacturing with deep learning, *Journal of Intelligent Manufacturing*, 31/8:2003–2017, https://doi.org/10.1007/s10845-020-01549-2.

[21] Aoyagi, K., Wang, H., Sudo, H., Chiba, A., 2019, Simple method to construct process maps for additive manufacturing using a support vector machine, *Additive Manufacturing*, 27:353–362, https://doi.org/10.1016/J.ADDMA.2019.03.013.

[22] Guo, L., Cheng, Y., Zhang, Y., Liu, Y., Wan, C., et al., 2021, Development of Cloud-Edge Collaborative Digital Twin System for FDM Additive Manufacturing, *IEEE International Conference on Industrial Informatics (INDIN)*, 2021-July/2018:1–6, DOI:10.1109/INDIN45523.2021.9557492.

[23] Wang, Y., Lin, Y., Zhong, R. Y., Xu, X., 2019, IoT-enabled cloud-based additive manufacturing platform to support rapid product development, *International Journal of Production Research*, 57/12:3975–3991, https://doi.org/10.1080/00207543.2018.1516905.

[24] Adnan, M., Lu, Y., Jones, A., Cheng, F. T., 2019, Application of the fog computing paradigm to additive manufacturing process monitoring and control, *Solid Freeform Fabrication 2019: Proceedings of the 30th Annual International Solid Freeform Fabrication Symposium - An Additive Manufacturing Conference*, SFF 2019, pp. 254–267, https://doi.org/10.2139/ssrn.3785854.

[25] Wu, H., Yu, Z., Wang, Y., 2019, Experimental study of the process failure diagnosis in additive manufacturing based on acoustic emission, *Measurement: Journal of the International Measurement Confederation*, 136:445–453, https://doi.org/10.1016/j.measurement.2018.12.067.

[26] Abhilash, P. M., Chakradhar, D., 2022, Image processing algorithm for detection, quantification and classification of microdefects in wire electric discharge machined precision finish cut surfaces, *Journal of Micromanufacturing*, 5/2:116–126, https://doi.org/10.1177/25165984211015410.

[27] Donegan, S. P., Schwalbach, E. J., Groeber, M. A., 2020, Zoning additive manufacturing process histories using unsupervised machine learning, *Materials Characterization*, 161:110123, https://doi.org/10.1016/j.matchar.2020.110123.

[28] Kulkarni, A., Bhatt, P. M., Kanyuck, A., Gupta, S. K., 2021, Using unsupervised learning for regulating deposition speed during robotic wire arc additive manufacturing, Proceedings of the ASME Design Engineering Technical Conference, 2:1–13, https://doi.org/10.1115/DETC2021-71865.

[29] Ren, W., Wen, G., Zhang, Z., Mazumder, J., 2022, Quality monitoring in additive manufacturing using emission spectroscopy and unsupervised deep learning, *Materials and Manufacturing Processes*, 37/11:1339–1346, https://doi.org/10.1080/10426914.2021.1906891.

[30] Panagiotis, S., Alexios, P., Vasilis, S., Dimitris, M., 2021, An AR based digital twin for laser based manufacturing process monitoring, *Procedia CIRP*, 102:258–263, https://doi.org/10.1016/j.procir.2021.09.044.

[31] Chen, Y., Peng, X., Kong, L., Dong, G., Remani, A., et al., 2021, Defect inspection technologies for additive manufacturing, *International Journal of Extreme Manufacturing*, 3/2:022002, https://doi.org/10.1088/2631-7990/abe0d0.

[32] Abhilash, P. M., Ahmed, A., 2023, An image - processing approach for polishing metal additive manufactured components to improve the dimensional accuracy and surface integrity, *The International Journal of Advanced Manufacturing Technology*, 125:1–21, https://doi.org/10.1007/s00170-023-10916-1.

[33] Cai, Y., Wang, Y., Burnett, M., 2020, Using augmented reality to build digital twin for reconfigurable additive manufacturing system, *Journal of Manufacturing Systems*, 56/April:598–604, https://doi.org/10.1016/j.jmsy.2020.04.005.

[34] Tao, W., Lai, Z. H., Leu, M. C., Yin, Z., Qin, R., 2019, A self-aware and active-guiding training & assistant system for worker-centered intelligent manufacturing, *Manufacturing Letters*, 21:45–49, https://doi.org/10.1016/j.mfglet.2019.08.003.

[35] Boban, J., Ahmed, A., Jithinraj, E. K., Rahman, M. A., Rahman, M., 2022, *Polishing of Additive Manufactured Metallic Components: Retrospect on Existing Methods and Future Prospects.* Springer London.

[36] Boban, J., Ahmed, A., 2022, Electric discharge assisted post-processing performance of high strength-to-weight ratio alloys fabricated using metal additive manufacturing, *CIRP Journal of Manufacturing Science and Technology*, 39:159–174, https://doi.org/10.1016/j.cirpj.2022.08.002.

[37] Bellalouna, F., 2021, Case study for design optimization using the digital twin approach, *Procedia CIRP*, 100:595–600, https://doi.org/10.1016/J.PROCIR.2021.05.129.

[38] Wang, Z., Jiang, C., Liu, P., Yang, W., Zhao, Y., et al., 2020, Uncertainty quantification and reduction in metal additive manufacturing, *Computational Materials*, 6/1:1–10, https://doi.org/10.1038/s41524-020-00444-x.

[39] Jacoby, M., Usländer, T., 2020, Digital twin and internet of things-current standards landscape, *Applied Sciences*, 10:6519, https://doi.org/10.3390/APP10186519.

[40] Wu, Y., Zhang, K., Zhang, Y., 2021, Digital twin networks: a survey, *IEEE Internet of Things Journal*, 8/18:13789–13804, https://doi.org/10.1109/JIOT.2021.3079510.

[41] Coorey, G., Figtree, G. A., Fletcher, D. F., Snelson, V. J., Vernon, S. T., et al., 2022, The health digital twin to tackle cardiovascular disease-a review of an emerging interdisciplinary field, *Digital Medicine*, 5/1:1–12, https://doi.org/10.1038/s41746-022-00640-7.

[42] Colosimo, B. M., Centofanti, F., Centofanti, F., 2022, Model interpretability, explainability and trust for manufacturing 4.0. In Antonio Lepore, Biagio Palumbo, Jean-Michel Poggi (Eds.), *Interpretability for industry 4.0 , Statistical and Machine Learning Approaches,* pp. 21–36. Springer Nature, Cham. https://doi.org/10.1007/978-3-031-12402-0_2.

[43] Lee, J. A., Sagong, M. J., Jung, J., Kim, E. S., Kim, H. S., 2023, Explainable machine learning for understanding and predicting geometry and defect types in Fe-Ni alloys fabricated by laser metal deposition additive manufacturing, *Journal of Materials Research and Technology*, 22:413–423, https://doi.org/10.1016/J.JMRT.2022.11.137.

[44] Shao, G., Helu, M., 2020, Framework for a digital twin in manufacturing: scope and requirements, *Manufacturing Letters*, 24:105–107, https://doi.org/10.1016/j.mfglet.2020.04.004.

[45] Löcklin, A., Jung, T., Jazdi, N., Ruppert, T., Weyrich, M., 2021, Architecture of a human-digital twin as common interface for operator 4.0 applications, *Procedia CIRP*, 104:458–463, https://doi.org/10.1016/J.PROCIR.2021.11.077.

Chapter 13

Additive manufacturing for society

Alex Y, Nidhin Divakaran, and Smita Mohanty
Central Institute of Petrochemicals
Engineering & Technology (CIPET)

13.1 INTRODUCTION

The rapid emergence and evolution of additive manufacturing (AM) (3D printing) has led to the creation of new and innovative systems for various industries. Initially, AM techniques were used for prototyping purposes. Now, these technologies have been widely used for the production and design of end products in various fields such as aerospace, automotive, medical, military/defense and industrial electronic products [1–3]. Figure 13.1 represents the supply of AM in different applications. Some of its advantages over conventional technologies, like being capable of producing sophisticated and customized products, made AM more attractive in this decade. Due to the technological advancements that have occurred, the quality and price of these products have significantly changed. However, it has recently seen a surge in popularity due to advances in technology, such as the use of metal, plastic, and ceramic materials [4,5].

The rapid growth of the AM industry has been attributed to the increasing number of developments in all fields, which has also enhanced research to its peak. In 2016, the global AM industry's revenues reached over 6.063 billion dollars. This represents a growth of 17.4% over the previous years. According to many experts [7–9], the rapid emergence and growth of the AM industry have identified it as a potential revolutionary technology that could transform the way the world works. The development of new technologies within the AM industry has the potential to affect various domains such as industrial, business, and society [10]. Though it has the potential to dramatically reduce the amount of time and resources needed to produce parts, which can have a profound impact on society.

The current state of AM is still in its developing stages [12]. There are still many challenges that need to be addressed, such as the accuracy of the parts produced, the cost of materials and machines, and the scalability of the process [13–15]. In addition, there are still significant barriers to entry, such as the lack of availability of high-quality materials and machines. Figure 13.2 displays the evolution of 3D printing technologies over the years. Despite these challenges, AM has the potential to revolutionize the way we design

DOI: 10.1201/9781003406488-13

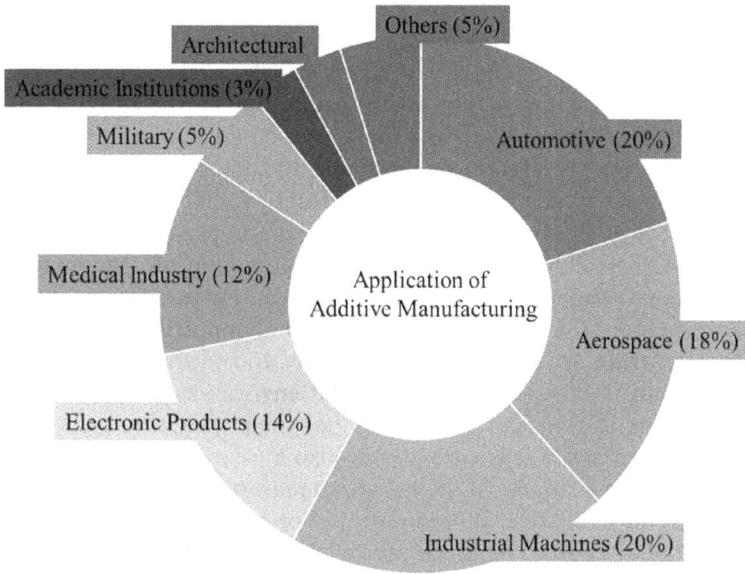

Figure 13.1 The supply of AM in different applications [6].

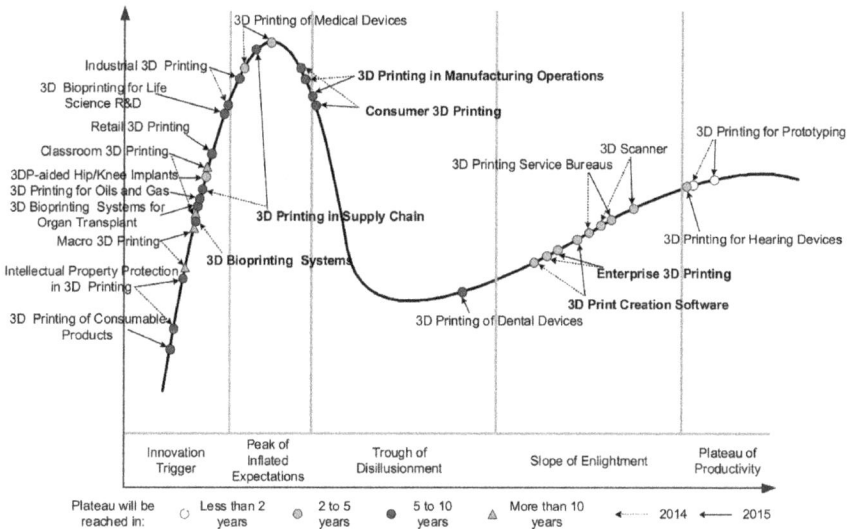

Figure 13.2 Evolution of 3D printing over the years (Reproduced with permission from [11]), Copyright Elsevier 2018.

and manufacture products. With the ability to create complex, customized parts in a fraction of the time, it is easier to deliver customer-satisfied products at minimal cost. Another important one is the cost of traditional manufacturing methods over new technology. It has the potential to drastically reduce the amount of resources needed to produce parts, which helps reduce the cost of production. Moreover, it helps reduce the energy consumption of the manufacturing process and improve the lead times. This could have a profound impact on society, as it could lead to faster, cheaper, and more efficient manufacturing processes. In addition, AM has the potential to create new products from custom materials that were not previously available [16,17]. By using new materials such as plastics and their composites, metals and their alloys, and the combination of composite alloys in AM, we can create products that are stronger, lighter, and more durable than ever before [18]. This could open up new opportunities in a variety of industries, such as aerospace, automotive, and medical [19]. However, it is important to consider the ethical and environmental implications of this technology [20]. In particular, AM has the potential to create custom products that are more efficient but may also have a negative impact on the environment. For example, 3D printing can require the use of plastic, which can be difficult to recycle. In addition, 3D printing can produce products that are more difficult to repair or reuse, which could lead to an increase in waste and pollution [21]. So, it is important to consider these implications when using AM to create products and to ensure that they are designed to be as sustainable as possible [22].

Due to increasing concerns about the environment and the need to minimize the impact of manufacturing on the planet, the adoption of sustainable practices has become more prevalent [23]. As mentioned earlier, AM technologies are regarded as sustainable due to their ability to reduce the amount of waste produced and improve the circular economy [24]. A literature review conducted on the life cycle assessment of AM revealed that it can produce savings by reducing material handling and supply chain complexities [25]. AM is revolutionizing the way products and parts are made. By using 3D printing technology, companies and individuals are able to reduce costs and improve time to market. Also, this AM technology has the potential to revolutionize society in a variety of ways. In this chapter, we will explore the potential impact of AM on society in different sectors. Also, we will examine the challenges associated with the adoption of AM as well as the potential for further development. Finally, we will discuss the importance of research and development in order to ensure that the benefits of AM are realized.

13.2 THE IMPACT OF ADDITIVE MANUFACTURING ON SOCIETY

AM is having an intense impact on society. It is revolutionizing the manufacturing process, making it faster, cheaper, and more efficient. It is also allowing for the creation of complex and intricate objects that were not previously possible. One of the most significant impacts of AM is its potential to revolutionize the medical industry [26]. 3D printing technology is being used to create medical implants, prosthetics, and even artificial organs. This could have a profound impact on the quality of life of those who need these devices. Another major impact of 3D printing is its potential to revolutionize the manufacturing industry [26]. By using 3D printing to create complex parts with intricate details, manufacturers can reduce production costs and speed up the manufacturing process. This could lead to more efficient production and the creation of new products. Finally, 3D printing is also having an impact on the environment. By reducing the amount of material waste in the manufacturing process, 3D printing helps to reduce the amount of pollution and energy consumption associated with traditional manufacturing processes [27].

One of the most significant benefits of AM is the ability to reduce waste. In traditional manufacturing processes, a great deal of material is wasted during the manufacturing process. This is because traditional manufacturing requires large amounts of material to be cut, drilled, and milled, resulting in a great deal of material waste [16]. With AM, objects can be printed layer by layer, using only the exact amount of material necessary. This can result in significant reductions in material waste, helping to reduce the environmental impact of manufacturing. Another significant benefit is that AM can also be used to create customized and personalized products [28]. In traditional manufacturing processes, the production of customized products is often expensive and time-consuming. With AM, customized products can be designed and produced quickly and cost-effectively. This means that customers can have access to products that are tailored to meet their exact needs [29]. In addition, AM can help increase the availability of medical devices and prosthetics in an emergency situation. By using 3D printing technology, medical device manufacturers can quickly and cost-effectively create customized prosthetics and medical devices with the help of the patient CT scan data. This means that people with disabilities or with critical medical conditions can have access to the devices they need faster and more affordably.

AM also has the potential to revolutionize the way products are made. By using 3D printing technology, companies can produce products with complex geometries and intricate designs that would otherwise be impossible to produce with traditional manufacturing methods. This can result in lighter and stronger products, as well as products with improved performance characteristics. The different parameters available within the 3D printing technology will help to modify the infill density from 10% to 100% (hollow-solid structure). This helps the creator manufacture a number of products with respect to the customer's needs without material loss. Finally, AM has the potential to create new business opportunities. By using 3D printing technology, entrepreneurs can quickly and easily create customized products for their customers. This can result in new and innovative products, as well as new business models and revenue streams.

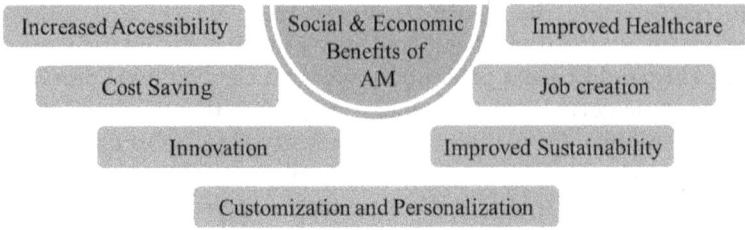

Figure 13.3 Social and economic benefits of AM technology.

In conclusion, AM has the potential to revolutionize society in a variety of ways. By reducing waste, providing access to customized products, and increasing the availability of medical devices and prosthetics, AM can help to improve the lives of individuals and communities. In addition, AM can help reduce the cost of product development and create new business opportunities [30]. As technology continues to improve, the potential for AM to benefit society is only going to increase. AM has brought many social and economic benefits in this decade, as displayed in Figure 13.3, some of which include the following:

1. Increased accessibility: 3D printing has made it possible to produce products and components in remote or rural areas, providing greater access to goods and services.
2. Improved sustainability: 3D printing can significantly reduce waste and energy consumption compared to traditional manufacturing methods, making it a more environmentally friendly option.
3. Customization and personalization: With 3D printing, it's possible to create customized products tailored to individual needs, enabling consumers to have more control over the products they purchase.
4. Job creation: The growth of the 3D printing industry is creating new job opportunities in areas such as product design, engineering, and advanced manufacturing.
5. Cost savings: 3D printing can reduce the cost of production, especially for low-volume or one-off products, by eliminating the need for expensive tooling and molding.
6. Improved healthcare: AM is enabling the production of customized prosthetics, implants, and surgical instruments, which can improve patient outcomes and quality of life.
7. Innovation: 3D printing is encouraging innovation and new product development, leading to the creation of new products and industries.

These benefits are transforming many aspects of society and the economy, and it is likely that the impact of AM will only continue to grow in the coming years.

13.3 NEED FOR THE STUDY OF ADDITIVE MANUFACTURING APPLICATIONS TOWARD ENVIRONMENTAL SUSTAINABILITY

The increasing need for sustainable practices in manufacturing has resulted in a requirement for new and innovative solutions. Using AM, companies can now achieve their goals (Figure 13.4). Through the use of AM, companies can now produce products that are both innovative and sustainable. It allows them to create new designs and reduce waste. Unlike traditional manufacturing techniques, which involve using rigid and hard materials, AM eliminates these disadvantages by using materials that are flexible and can be printed. It also significantly reduces waste. The high degree of automation and control that 3D printing software provides allows it to achieve accuracy levels that are typically impossible to achieve with other technologies. The ability to control the entire process of 3D printing allows it to reduce the time it takes to transfer data. This eliminates the need for manual data entry and allows designers to create faster concepts. A 3D printer can create a component without tooling, which eliminates the need for manual

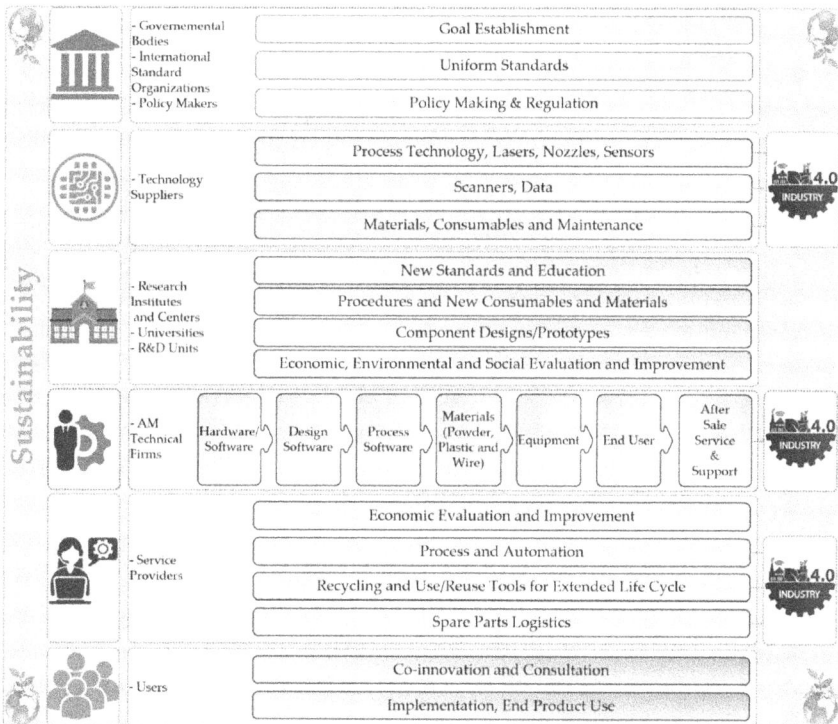

Figure 13.4 The common decision undertaken by stakeholders toward contribution of AM toward environmental sustainability in industries. (Reproduced with permission from [9], Copyright MDPI 2020.)

assembly and significantly lowers the time it takes to produce parts. Since most manufacturing processes are automated, there is no need for manual intervention, which helps lower labor costs and encourages the hiring of less skilled workers. The ability to create a product design flaw in different manufacturing methods can be easily addressed with AM, which eliminates the need for significant changes in the process. The possible causes of this issue include part orientation, material quality, and system calibration. To ensure that the parts are compatible, the fabrication parameters of the AM components should be monitored and analyzed.

13.4 APPLICATION OF ADDITIVE MANUFACTURING TECHNOLOGY IN SOCIETY

AM has the potential to revolutionize the entire manufacturing industry. It is capable of producing complex parts and components rapidly and with a high degree of accuracy. This technology can reduce the amount of time, labor, and resources needed to create products. It is also capable of producing parts that are lighter, stronger, and more efficient than those created using traditional manufacturing methods. The applications of AM technology are widely spread across numerous sectors and have become an increasingly common tool in many industries, especially in areas such as aerospace and defense, automotive, medical, dental, consumer products, and architecture. In the aerospace industry, 3D printing is being used to create lighter, stronger airplane parts and components, such as engine nozzles, fuselage components, and airframes. In healthcare, 3D printing can be used to create customized medical implants, prosthetics, hearing aids, and medical equipment, such as ventilators and surgical tools, in a more efficient and cost-effective manner [31]. Additionally, 3D printing can be used to create personalized educational materials and tools, such as interactive models, for students from schools to colleges. Several schools all across the world, including in secondary and primary schools. It allows educators to provide their students with hands-on learning opportunities that are beneficial to their scientific understanding [32]. Today, technical and third-level universities have started incorporating 3D printing projects and modules into their engineering courses [33]. In the environmental sector, AM can be used to create products with a smaller carbon footprint. For example, 3D printing can be used to create components for wind turbines, solar panels, and other renewable energy sources [34]. In the automotive industry, 3D printing is being used to create custom car parts, such as steering wheels, gearboxes, and brake pads [35]. Additionally, 3D printing is being used in the consumer product industry, such as to create custom jewelry and consumer electronics. Finally, in the architecture industry, 3D printing is being used to create full-scale 3D models of buildings, bridges, and other structures. All these evidences clearly show that the manufactures and customers adopted AM technology for sustainable development [23].

13.4.1 Application of additive manufacturing technology in aerospace industry

AM technology is becoming increasingly important in the aerospace industry. This technology can be used to create complex parts for aircraft, allowing for lighter, stronger, and more efficient designs for aircraft and spacecraft. It can also be used to quickly and cost-effectively produce replacement parts for aircraft, reducing maintenance costs and downtime. AM is also being used to produce tools and fixtures for the aerospace industry, enabling faster and more accurate production processes. In addition, 3D printing can be used to produce prototypes and mock-ups of components and systems, allowing engineers to quickly test and refine designs. Finally, AM technology is being used to produce various parts for both space and earth-bound applications, allowing for more efficient designs and faster production processes. Recently, AM has been used to create complex structures such as engines and wings [17]. This technology can reduce the amount of time, resources, and energy needed to create aircraft and spacecraft.

The AM system allows designers to create advanced parts that are designed to provide additional space and better performance. These include multi-functional parts, parts that are hard to machine, and parts that are designed to consolidate. The ability to create freeform designs using the AM system makes it an ideal tool for the aerospace industry. Since it was first introduced, several aerospace companies, including Boeing, have used the system to produce thousands of parts. In order to reduce its costs, Boeing has started to produce titanium AM components. These parts will allow the company to save up to 3 million dollars per aircraft. GE Aviation is one of the companies that uses metal AM to produce thousands of fuel nozzles for its LEAP engine. Airbus is also a user of the system. Through its partnership with Arconic, Boeing will be able to produce large-scale AM components [36].

13.4.2 Application of additive manufacturing technology in defense

AM technology has been making a major impact in the military/defense sectors. This technology has been used to reduce production costs, reduce lead times, and improve the performance of defense systems during emergency times. AM also gives the military the ability to quickly produce complex, custom parts that would otherwise be impossible to produce with traditional manufacturing techniques [37]. AM has been used to create components for aircraft, tanks, and other weapons systems, such as missile launchers, artillery pieces, and robots. This technology has also been used to create components for military vehicles, such as armor plating, engine parts, and other components that require precision and strength. In addition, AM has enabled the production of parts for drones, which are becoming increasingly important in the theater of operations. AM has also been used in the development of protective equipment, such as helmets, body armor, and other protective gear. This technology has allowed the military to create custom-fitted gear that can better protect soldiers from the dangers of combat. This technology has allowed for the creation of lightweight and custom-fitted medical items that can be used to treat wounded soldiers on the battlefield.

13.4.3 Application of additive manufacturing technology in automotive industry

AM technology is increasingly being used in the automotive industry for the production of automotive parts. Advantages of AM technology over traditional manufacturing techniques include faster prototyping, improved product design, reduced material waste, and the ability to create parts with complex geometries [6]. Examples of automotive parts that can be made using AM technology include fuel injectors, air intake manifolds, and engine blocks [38]. 3D printing can also be used for the production of customized car parts for restoration projects or vintage car restorations. In addition, 3D printing can be used to produce lighter, stronger, and more efficient parts, such as turbochargers, exhaust manifolds, and intake systems, which can improve fuel efficiency and reduce emissions. It can be used to create custom parts and components for cars, trucks, and other vehicles. This technology can also be used to create custom 3D printed car bodies and frames. Additionally, AM can be used to create custom interior components, such as dashboards and consoles. Finally, 3D printing can be used to produce replacement parts for cars that are no longer in production, thereby reducing costs and the need to purchase new parts.

13.4.4 Application of additive manufacturing technology in medical research

AM is a rapidly emerging technology that has the potential to revolutionize medical research and is being used in the healthcare and biomedical engineering fields. AM technology has the potential to create complex, personalized medical devices, tissue engineering, regenerative medicine, and equipment that are more accurate and efficient than ever before [39]. It can also be used to produce patient-specific 3D models for pre-surgical planning [40,41]. Additionally, AM technology can be used to fabricate patient-specific tissue scaffolds that can be used in regenerative medicine and tissue engineering applications [42]. AM can also be used to create personalized medical implants, such as bone and joint replacements, that are tailored to the patient's anatomy, allowing for a better fit and improved outcomes. Finally, AM technology can be used to create custom-made prosthetics that are more comfortable and lifelike than traditional prosthetics. By leveraging the potential of AM technology, medical researchers are able to develop effective, personalized treatments for a variety of medical conditions, especially in drug delivery systems, enabling the production of personalized, patient-specific dosage forms. 3D printing can also be used for tissue engineering and regenerative medicine, allowing for the production of customized scaffolds and tissue structures. Finally, 3D printing can be used for personalized medicine, allowing for the production of tailored medical treatments [43]. Additionally, AM can be used to produce custom parts for medical devices and equipment, such as replacement parts for MRI machines.

13.4.4.1 Application of additive manufacturing technology in orthopedics

AM technology has become an important tool in orthopedic prosthetics and devices, including hearing aids and orthodontic braces. It is being used to create customized implants, prosthetics, and orthoses [44]. It also allows for the production of lightweight components with increased strength and durability. In addition, the use of 3D printing allows for the production of customized implants that are tailored to the patient's exact anatomy, thus reducing the risk of implant rejection. AM technology is also being used to produce patient-specific prosthetic devices, such as orthoses for lower-limb amputees [45]. Finally, AM technology is being used to create custom tools and fixtures for use in orthopedic surgeries [46].

13.4.4.2 Application of additive manufacturing technology in dentistry

AM technology has a wide range of applications in dentistry. It can be used to produce dental prosthesis, crowns, bridges and implants. It can also be used for producing custom dental appliances, such as orthodontic retainers, nightguards and bleaching trays [47]. Additionally, AM technology can be used to create models for planning complex dental treatments, such as implant placement. In the future, it may also be used to produce 3D-printed dental implants and scaffolds for tissue regeneration [48].

13.4.5 Application of additive manufacturing technology in consumer products

AM technology has been applied to several consumer products. One example is the 3D printing of customized smartphone cases. The technology has made it possible to create a wide variety of designs and materials, allowing consumers to customize their devices to their own tastes [49]. Other consumer products that have been created using AM technology include jewelry, apparel, eyewear, and toys. With further development of the technology, it is expected that it will be used to produce more complex consumer products such as furniture and home décor [50]. Later, Xiong observes [51] the materials, application status, and development trends of AM technology. He discusses the materials used in AM and their potential applications, including their use in aerospace, medical, automotive, and other industries. Xiong then reviews the development progress of various AM technologies, such as selective laser sintering, 3D printing, and electron beam melting. He then concludes by examining the development trends of AM technology, including the development of new materials, new applications, and automation [51].

AM technology has the potential to revolutionize construction and architecture. In fact, it has already been used in the construction of a range of projects, such as bridges, houses, and even entire skyscrapers [52]. AM technology offers architects a wide range of opportunities to create complex and intricate designs and structures that would be impossible to create with traditional building methods. For instance, with AM, it is possible to create structures with intricate internal networks of tunnels, pipes, and other elements that would be difficult to construct with traditional methods. AM technology also has the potential to reduce the amount of material needed to construct a given project, which can help reduce costs and environmental impacts. Additionally, the technology can be used to create intricate and complex shapes that cannot be achieved with traditional building approaches. Finally, AM technology can also speed up the construction process, as the components of a given project can be printed on-site, leading to a faster and more efficient construction process. This has the potential to reduce the amount of labor needed for a project as well as the time required for a project to be completed. Li et al. [53] discuss the development of a new architecture design for periodic truss-lattice cells for AM, which offers a new possibility for lightweight and high-performance structures. The design includes a periodic truss-lattice unit cell and a topology optimization algorithm. The structure is optimized by introducing a new design variable that incorporates the size and shape of the unit cell as well as its truss-lattice arrangement. Using this design, the authors were able to achieve lightweight and high-performance designs with equal or better performance than existing designs. The results demonstrate the potential of this new architecture for achieving lightweight and high-performance structures for AM [53]. The implementation of a robot control architecture for use in AM applications [54]. The architecture is based on a hierarchical structure, with a global controller at the highest level and a local controller at the lowest level. The global controller is responsible for managing the overall task, and the local controller is responsible for managing the individual robot commands. The architecture is tested on a 3D printer with an industrial robotic arm. The results demonstrate that the architecture is capable of providing accurate and repeatable performance, as well as being robust to environmental disturbances. This suggests that the proposed architecture could be used for a variety of AM applications [54]. AM can be used to create components for buildings, such as walls, columns, and beams, that are lightweight, strong, and customized to the exact specifications of the project. This means that construction projects can be completed faster and with fewer materials, reducing costs and waste. Additionally, 3D printing can be used to create durable and complex geometries that would be difficult to achieve with traditional construction procedures. This could open up new possibilities for the design and construction of buildings, bridges, and other structures.

AM technology can be applied in machine tool production to reduce cost, optimize production cycles, and improve the performance of machine tools. 3D printing is increasingly being used to produce parts and components for machine tools, as it offers a range of advantages over traditional manufacturing techniques. Structural components such as brackets, housings, and covers can be produced in one piece, reducing the need for welding and assembly [55]. This reduces the cost and time associated with traditional manufacturing processes. Furthermore, 3D printing can be used to produce custom parts tailored to the specific requirements of a machine tool. This can result in improved performance and reliability. In addition, 3D printing can be used to rapidly prototype new parts, allowing for faster development cycles [50]. This can help reduce the time to market for machine tools.

13.4.6 Application of additive manufacturing technology in fashion and jewelry

AM is an emerging technology that has the potential to revolutionize the jewelry industry. AM has the potential to reduce production costs, optimize production processes, and open up new possibilities in design. With the help of 3D printing, jewelry designers can create unique and intricate designs with a greater level of detail and complexity than ever before [56]. Additionally, 3D printing eliminates the need for traditional casting and polishing operations, reducing production time and cost. This technology can also be used to produce customized jewelry at a fraction of the cost of traditional methods. In addition, AM can be used for rapid prototyping and rapid tooling, allowing for shorter product development cycles and a faster time to market [57]. Finally, AM can be used to produce complex geometries, such as organic shapes and intricate patterns, that are difficult to replicate with traditional manufacturing methods [58].

AM technology is increasingly being applied to fashion. 3D printing of fabrics, garments, and accessories has become an attractive option for fashion designers as it allows them to create complex shapes, intricate details, and unique textures with ease. This technology provides fashion designers with the opportunity to create one-of-a-kind pieces that stand out from the crowd. 3D-printed clothing and accessories can be customized to fit the wearer perfectly, making it an attractive option for those looking for personalization [59]. Additionally, the use of 3D printing technology in fashion can lead to a more sustainable approach to fashion production by reducing waste, cutting down on lead times for production, and eliminating the need for costly molds and dies.

13.4.7 Application of additive manufacturing technology in food industry

AM technology can be applied in the food industry to improve production processes, reduce costs, and improve product quality. AM can be used to manufacture custom-made food products, such as cookies, cakes, and breads, with intricate details that could not be achieved with traditional manufacturing techniques. 3D printing could also be used to create complex shapes, such as molds for chocolate, that are more aesthetically pleasing than traditional molds. AM could also be used to create food packaging with intricate designs that could be more appealing to customers [60]. Furthermore, AM could provide a more efficient way to package food items, as custom-made packaging that fits the product better could be created [61]. Additionally, AM could also be used to create food-grade materials, such as edible films or coatings, that could better preserve food products and extend their shelf life.

13.4.8 Application of additive manufacturing technology in entertainment fields

AM technology is being increasingly applied in the entertainment industry, from film production to video games. 3D printing, for example, can be used to create accurate scale models for use in films and television and can even be used to create props and costumes. Video game designers are also leveraging 3D printing technology to create high-fidelity models of characters and environments, as well as limited-edition figurines and game pieces. 3D printing can also be used to create custom items such as game controllers and accessories, allowing gamers to customize their gaming experience. Additionally, 3D printing can be used to create amusement park rides and attractions. By using 3D printing, manufacturers can quickly and cost-effectively create intricate, one-of-a-kind designs.

13.5 CHALLENGES AND FUTURE OF ADDITIVE MANUFACTURING FOR SOCIETY

AM has the potential to revolutionize the way society manufactures goods, from consumer products to numerous day-to-day using devices. With the rise of 3D printing, the cost of production for many products is reduced, making them more accessible to consumers. Additionally, 3D printing offers more customization options for consumers, allowing them to design their own products. This could lead to a more personalized consumer experience. It has the potential to create new job opportunities in the manufacturing industry as more companies look to utilize 3D printing for production. This could lead to an increase in skilled labor in the manufacturing industry as well as an increase in new technology jobs. Furthermore, 3D printing could potentially reduce the amount of waste generated during manufacturing processes. By utilizing 3D printing, companies could reduce or eliminate the need for large amounts of plastic or other materials to produce a product. This could lead to a more sustainable and environmentally friendly manufacturing process. Overall, the potential for AM for society is immense and could lead to a more efficient and cost-effective manufacturing process, new job opportunities, a reduction in waste, and improved medical treatments. AM is changing the world in a number of ways, including:

1. Revolutionizing manufacturing: 3D printing is transforming the way products are designed, manufactured, and distributed, allowing for more customization and on-demand production.
2. Enabling healthcare advances: 3D printing is being used to create customized prosthetics, implants, and surgical instruments, improving patient outcomes and quality of life.
3. Encouraging innovation: 3D printing is enabling innovation in product design and development, leading to the creation of new products and industries.
4. Improving sustainability: 3D printing can significantly reduce waste and energy consumption compared to traditional manufacturing methods, making it a more environmentally friendly option.
5. Democratizing access: 3D printing is making it possible to produce products and components in remote or rural areas, providing greater access to goods and services.
6. Boosting the economy: The growth of the 3D printing industry is creating new job opportunities and contributing to the economy.
7. Transforming education: 3D printing is being used in education to provide hands-on learning experiences, encouraging students to develop critical thinking and problem-solving skills.

Overall, AM is changing the way we live and work, and it will likely continue to have a significant impact in the future. It is also being used to help address the global challenges of climate change and resource scarcity. Overall, AM has had a positive impact on society, offering a more efficient and sustainable way to manufacture products. It can also be used to make prototypes of new products, allowing companies to quickly test ideas and products before committing to a large-scale manufacturing run. AM is a highly useful technology that can benefit a variety of industries and society as a whole. It can be used to create customized, complex parts without the need for traditional machining and tooling. This can reduce production costs and lead times while also allowing for more efficient use of materials and energy. In addition, AM can help reduce waste and allow companies to produce parts on demand, which can reduce inventory costs. AM can also be used to create parts and products that are not economically feasible to produce using traditional manufacturing techniques. Finally, it can be used to create parts that would be impossible to create using traditional methods, such as parts with complex internal geometries.

All of these benefits can lead to a more efficient, cost-effective, and sustainable manufacturing process, which can benefit society as a whole. Figure 13.5 shows the various strategies that are utilized by AM to drive sustainability. The various aspects of sustainability are addressed in this

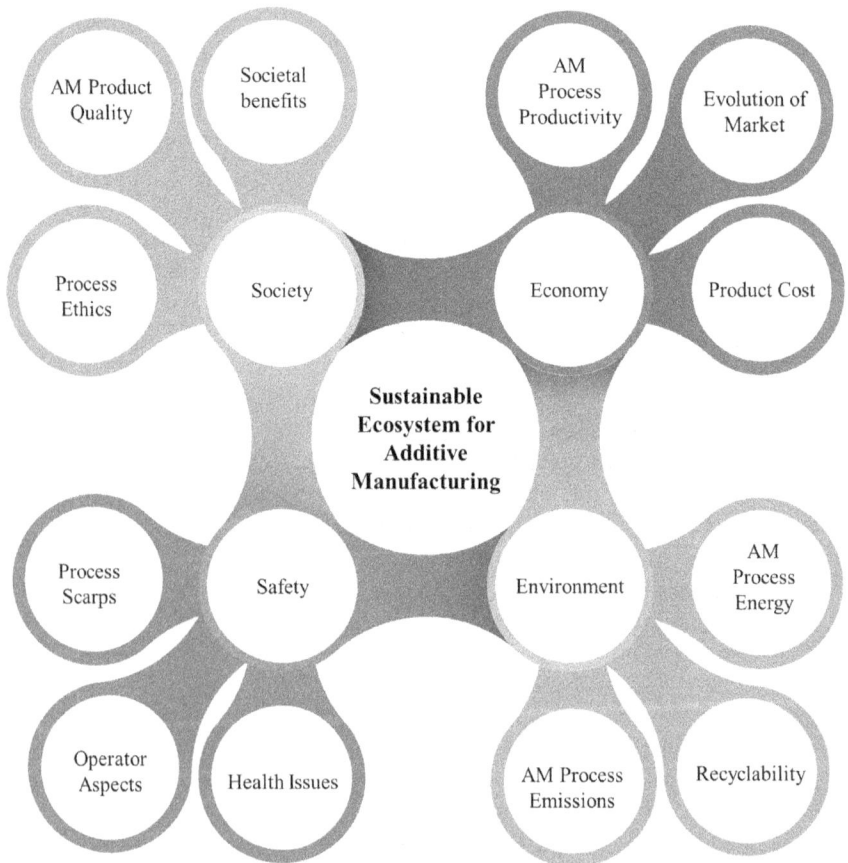

Figure 13.5 The various strategies that are utilized by AM to drive sustainability. Source Referred from [62].

section, such as the safety aspect and the economic and environmental benefits of AM. In addition, the process energy and emissions are covered to make the technology completely sustainable. Despite its many benefits, AM also faces several challenges and limitations, some of which include:

1. Material limitations: The selection of materials that can be used in 3D printing is currently limited, and there is a need for new and improved materials with specific properties and characteristics.
2. Quality and accuracy: The quality and accuracy of 3D printed parts can be affected by factors such as layer thickness, material properties, and machine calibration, making it necessary to continually improve and refine 3D printing technology.
3. Speed: 3D printing can be slow compared to traditional manufacturing methods, making it unsuitable for mass production applications.
4. Cost: While 3D printing can reduce costs for low-volume or one-off products, the cost of 3D printers and materials is still high, making it an expensive option for some applications.
5. Intellectual property: 3D printing raises concerns about intellectual property protection and piracy, as it is easier to create and distribute copies of products.
6. Regulation: There are currently few regulations in place to ensure the safety and quality of 3D printed products, making it necessary to develop and enforce new standards.
7. Public perception: There is a lack of understanding and awareness about 3D printing among the general public, which can lead to misconceptions and resistance to its widespread adoption.

These challenges and limitations will need to be addressed in order for 3D printing to reach its full potential and continue to bring benefits to society. However, with continued research and development, it is likely that many of these challenges will be overcome in the coming years.

13.6 CONCLUSION

AM has the potential to transform the way we design and manufacture products. With the ability to produce complex parts with a fraction of the time and cost of traditional manufacturing methods, AM has the potential to drastically reduce the amount of time and resources needed to produce parts. However, it is important to consider the ethical and environmental implications of this technology and to ensure that the products created are as sustainable as possible. By doing so, AM could be used to create a more sustainable society.

AM is having a profound impact on society. It is revolutionizing the medical industry, the manufacturing industry, and the environment. Its potential to create complex and intricate objects opens up new possibilities for product design and manufacturing. As technology continues to evolve, its impact on society will only increase.

In this book chapter, "Additive Manufacturing for Society," we explored the impact that this technology is having on society as a whole. We discussed and included the various applications of AM in industries such as healthcare, aerospace, and consumer goods and examined the social and economic benefits it can bring. Additionally, we will also look at some of the challenges and limitations that need to be addressed in order for AM to reach its full potential. This chapter will provide a comprehensive overview of how AM is changing our world and what the future holds for this exciting and rapidly evolving field.

REFERENCES

[1] M. Attaran, The rise of 3-D printing: the advantages of additive manufacturing over traditional manufacturing, *Business Horizons.* 60 (2017) 677–688. https://doi.org/10.1016/j.bushor.2017.05.011.

[2] H. Dodziuk, Applications of 3D printing in healthcare, Kardiochirurgia i Torakochirurgia Polska. 13 (2016) 283–293. https://doi.org/10.5114/kitp.2016.62625.

[3] B. Berman, 3-D printing: the new industrial revolution, *Business Horizons.* 55 (2012) 155–162. https://doi.org/10.1016/j.bushor.2011.11.003.

[4] A.M. Schauer, K.B. Fillingim, K. Fu, Impact of timing in the design process on students' application of design for additive manufacturing heuristics, *Journal of Mechanical Design, Transactions of the ASME.* 144 (2022) 12. https://doi.org/10.1115/1.4053281.

[5] K.S. Prakash, T. Nancharaih, V.V.S. Rao, *Additive Manufacturing Techniques in Manufacturing -An Overview, Materials Today Proceeding.* (2018). https://doi.org/10.1016/j.matpr.2017.11.642.

[6] A. Vafadar, F. Guzzomi, A. Rassau, K. Hayward, Advances in metal additive manufacturing: a review of common processes, *Industrial Applications, and Current Challenges, Applied Sciences.* 11 (2021) 1213. https://doi.org/10.3390/app11031213.

[7] M. Xu, J.M. David, S.H. Kim, The fourth industrial revolution: opportunities and challenges, *International Journal of Financial Research.* 9 (2018) 90. https://doi.org/10.5430/ijfr.v9n2p90.

[8] M.K. Niaki, F. Nonino, Impact of additive manufacturing on business competitiveness: a multiple case study, *Journal of Manufacturing Technology Management.* 28 (2017) 56–74. https://doi.org/10.1108/JMTM-01-2016-0001.

[9] R. Godina, I. Ribeiro, F. Matos, B.T. Ferreira, H. Carvalho, P. Peças, Impact assessment of additive manufacturing on sustainable business models in industry 4.0 context, *Sustainability (Switzerland).* 12 (2020) 7066. https://doi.org/10.3390/su12177066.

[10] F. Caviggioli, E. Ughetto, A bibliometric analysis of the research dealing with the impact of additive manufacturing on industry, business and society, *International Journal of Production Economics.* 208 (2019) 254–268. https://doi.org/10.1016/j.ijpe.2018.11.022.

[11] T. Peng, K. Kellens, R. Tang, C. Chen, G. Chen, Sustainability of additive manufacturing: an overview on its energy demand and environmental impact, *Additive Manufacturing.* 21 (2018) 694–704. https://doi.org/10.1016/j.addma.2018.04.022.

[12] M. Pérez, D. Carou, E.M. Rubio, R. Teti, Current advances in additive manufacturing, *Procedia CIRP.* 88 (2020) 439–444. https://doi.org/10.1016/j.procir.2020.05.076.

[13] Y.S. Leung, T.H. Kwok, X. Li, Y. Yang, C.C.L. Wang, Y. Chen, Challenges and status on design and computation for emerging additive manufacturing technologies, *Journal of Computing and Information Science in Engineering.* 19 (2019) 021013. https://doi.org/10.1115/1.4041913.

[14] R. He, N. Zhou, K. Zhang, X. Zhang, L. Zhang, W. Wang, D. Fang, Progress and challenges towards additive manufacturing of SiC ceramic, *Journal of Advanced Ceramics.* 10 (2021) 637–674. https://doi.org/10.1007/s40145-021-0484-z.

[15] L.Y. Chen, S.X. Liang, Y. Liu, L.C. Zhang, Additive manufacturing of metallic lattice structures: unconstrained design, accurate fabrication, fascinated performances, and challenges, *Materials Science and Engineering R: Reports.* 146 (2021) 100648. https://doi.org/10.1016/j.mser.2021.100648.

[16] T. Pereira, J.V. Kennedy, J. Potgieter, A comparison of traditional manufacturing vs additive manufacturing, the best method for the job, *Procedia Manufacturing.* 30 (2019) 11–18. https://doi.org/10.1016/j.promfg.2019.02.003.

[17] A. Gisario, M. Kazarian, F. Martina, M. Mehrpouya, Metal additive manufacturing in the commercial aviation industry: a review, *Journal of Manufacturing Systems.* 53 (2019) 124–149. https://doi.org/10.1016/j.jmsy.2019.08.005.

[18] K. Song, Y. Cui, T. Tao, X. Meng, M. Sone, M. Yoshino, S. Umezu, H. Sato, New metal-plastic hybrid additive manufacturing for precise fabrication of arbitrary metal patterns on external and even internal surfaces of 3D plastic structures, *ACS Applied Materials & Interfaces.* 14 (2022) 46896–46911. https://doi.org/10.1021/acsami.2c10617.

[19] R. Singh, A. Gupta, O. Tripathi, S. Srivastava, B. Singh, A. Awasthi, S.K. Rajput, P. Sonia, P. Singhal, K.K. Saxena, Powder bed fusion process in additive manufacturing: an overview, *Materials Today Proceeding.* 30 (2019) 3058–3070. https://doi.org/10.1016/j.matpr.2020.02.635.

[20] F. Matos, C. Jacinto, Additive manufacturing technology: mapping social impacts, *Journal of Manufacturing Technology Management.* 30 (2019). https://doi.org/10.1108/JMTM-12-2017-0263.

[21] F.A. Cruz Sanchez, H. Boudaoud, M. Camargo, J.M. Pearce, Plastic recycling in additive manufacturing: a systematic literature review and opportunities for the circular economy, *Journal of Cleaner Production.* 264 (2020) 121602. https://doi.org/10.1016/j.jclepro.2020.121602.

[22] H.A. Colorado, E.I.G. Velásquez, S.N. Monteiro, Sustainability of additive manufacturing: the circular economy of materials and environmental perspectives, *Journal of Materials Research and Technology.* 9 (2020)) 8221–8234. https://doi.org/10.1016/j.jmrt.2020.04.062.

[23] M.K. Niaki, S.A. Torabi, F. Nonino, Why manufacturers adopt additive manufacturing technologies: the role of sustainability, *Journal of Cleaner Production.* 222 (2019) 381–392. https://doi.org/10.1016/j.jclepro.2019.03.019.

[24] J. Priyadarshini, R.K. Singh, R. Mishra, M.M Kamal, Adoption of additive manufacturing for sustainable operations in the era of circular economy: self-assessment framework with case illustration, *Computers & Industrial Engineering.* 171 (2022) 108514. https://doi.org/10.1016/j.cie.2022.108514.

[25] B. Naghshineh, F. Lourenço, R. Godina, C. Jacinto, H. Carvalho, A social life cycle assessment framework for additive manufacturing products, *Applied Sciences (Switzerland).* 10 (2020) 4459. https://doi.org/10.3390/app10134459.

[26] S. Rouf, A. Malik, N. Singh, A. Raina, N. Naveed, M.I.H. Siddiqui, M.I.U. Haq, Additive manufacturing technologies: industrial and medical applications, *Sustainable Operations and Computers.* 3 (2022) 258–274. https://doi.org/10.1016/j.susoc.2022.05.001.

[27] M. Shuaib, A. Haleem, S. Kumar, M. Javaid, Impact of 3D printing on the environment: a literature-based study, *Sustainable Operations and Computers.* 2 (2021) 57–63. https://doi.org/10.1016/j.susoc.2021.04.001.

[28] B.P. Conner, G.P. Manogharan, A.N. Martof, L.M. Rodomsky, C.M. Rodomsky, D.C. Jordan, J.W. Limperos, Making sense of 3-D printing: creating a map of additive manufacturing products and services, *Additive Manufacturing.* 1 (2014) 1–4. https://doi.org/10.1016/j.addma.2014.08.005.

[29] P. Reeves, C. Tuck, R. Hague, Additive manufacturing for mass customization, *Mass Customization.* (2011) 275–289. https://doi.org/10.1007/978-1-84996-489-0_13.

[30] I. Gibson, D. Rosen, B. Stucker, Business opportunities and future directions, In: *Additive Manufacturing Technologies: 3D Printing, Rapid Prototyping, and Direct Digital Manufacturing,* 475–486, Springer, New York, 2015. https://doi.org/10.1007/978-1-4939-2113-3_20.

[31] C. Lee Ventola, Medical applications for 3D printing: current and projected uses, *P and T.* 39 (2014) 704–711.

[32] D. Assante, G.M. Cennamo, L. Placidi, 3D printing in education: an european perspective, In: *IEEE Global Engineering Education Conference,* EDUCON, 2020. https://doi.org/10.1109/EDUCON45650.2020.9125311.

[33] S. Ford, T. Minshall, Invited review article: where and how 3D printing is used in teaching and education, *Additive Manufacturing.* 25 (2019) 131–150. https://doi.org/10.1016/j.addma.2018.10.028.

[34] M.N. Nadagouda, M. Ginn, V. Rastogi, A review of 3D printing techniques for environmental applications, *Current Opinion in Chemical Engineering.* 28 (2020) 173–178. https://doi.org/10.1016/j.coche.2020.08.002.

[35] M.R.C. Coimbra, T.P. Barbosa, C.M.A. Vasques, A 3D-printed continuously variable transmission for an electric vehicle prototype, *Machines.* 10 (2022) 84. https://doi.org/10.3390/machines10020084.

[36] J.C. Najmon, S. Raeisi, A. Tovar, Review of additive manufacturing technologies and applications in the aerospace industry, *Additive Manufacturing for the Aerospace Industry.* (2019) 7–31. https://doi.org/10.1016/B978-0-12-814062-8.00002-9.

[37] C. Brose, The new revolution in military affairs: war's sci-fi future, *Foreign Affairs.* 98 (2019).

[38] J. Gray, C. Depcik, Review of additive manufacturing for internal combustion engine components, *SAE International Journal of Engines.* 13 (2020) 617–632. https://doi.org/10.4271/03-13-05-0039.

[39] J. Witowski, N. Wake, A. Grochowska, Z. Sun, A. Budzyński, P. Major, T.J. Popiela, M. Pędziwiatr, Investigating accuracy of 3D printed liver models with computed tomography, *Quantitative Imaging in Medicine and Surgery.* 9 (2019) 43–52. https://doi.org/10.21037/qims.2018.09.16.

[40] C. Cooke, T. Flaxman, A. Sheikh, O. Miguel, M. McInnes, S. Singh, Pre-surgical planning using patient-specific 3D printed anatomical models for women with uterine fibroids, *Journal of Obstetrics and Gynaecology Canada.* 43 (2021) 670. https://doi.org/10.1016/j.jogc.2021.02.071.

[41] K.S. Hung, M.J. Paulsen, H. Wang, C. Hironaka, Y.J. Woo, Custom patient-specific three-dimensional printed mitral valve models for pre-operative patient education enhance patient satisfaction and understanding, *Journal of Medical Devices, Transactions of the ASME.* 13 (2019) 034501. https://doi.org/10.1115/1.4043737.

[42] S. Das, B. Basu, An overview of hydrogel-based bioinks for 3D bioprinting of soft tissues, *Journal of the Indian Institute of Science.* 99 (2019) 405–428. https://doi.org/10.1007/s41745-019-00129-5.

[43] S. Beg, W.H. Almalki, A. Malik, M. Farhan, M. Aatif, Z. Rahman, N.K. Alruwaili, M. Alrobaian, M. Tarique, M. Rahman, 3D printing for drug delivery and biomedical applications, *Drug Discovery Today.* 25 (2020) 1668–1681. https://doi.org/10.1016/j.drudis.2020.07.007.

[44] D.J. Thomas, D. Singh, 3D printing for developing patient specific cosmetic prosthetics at the point of care, *International Journal of Surgery.* 80 (2020) 36–39. https://doi.org/10.1016/j.ijsu.2020.04.023.

[45] G.R. Gubbala, R. Inala, Design and development of patient-specific prosthetic socket for lower limb amputation, *Material Science, Engineering and Applications.* 1 (2021) 7. https://doi.org/10.21595/msea.2021.22012.

[46] P. Andrés-Cano, J.A. Calvo-Haro, F. Fillat-Gomà, I. Andrés-Cano, R. Perez-Mañanes, Role of the orthopaedic surgeon in 3D printing: current applications and legal issues for a personalized medicine, *Revista Española de Cirugía Ortopédica y Traumatología.* 65 (2021) 138–151. https://doi.org/10.1016/j.recot.2020.06.014.

[47] M. Revilla-León, M. Sadeghpour, M. Özcan, A review of the applications of additive manufacturing technologies used to fabricate metals in implant dentistry, *Journal of Prosthodontics.* 29 (2020) 579–593. https://doi.org/10.1111/jopr.13212.

[48] M. Javaid, A. Haleem, Current status and applications of additive manufacturing in dentistry: a literature-based review, *Journal of Oral Biology and Craniofacial Research.* 9 (2019) 179–185. https://doi.org/10.1016/j.jobcr.2019.04.004.

[49] A. Haleem, M. Javaid, Additive manufacturing applications in industry 4.0: a review, *Journal of Industrial Integration and Management.* 04 (2019) 1930001. https://doi.org/10.1142/s2424862219300011.

[50] M. Mehrpouya, A. Dehghanghadikolaei, B. Fotovvati, A. Vosooghnia, S.S. Emamian, A. Gisario, The potential of additive manufacturing in the smart factory industrial 4.0: a review, *Applied Sciences (Switzerland).* 9 (2019) 3865. https://doi.org/10.3390/app9183865.

[51] S. Xiong, Materials, application status and development trends of additive manufacturing technology, *Materials Transactions.* 61 (2020) 1191–1199. https://doi.org/10.2320/matertrans.MT-M2020023.

[52] S.M.E. Sepasgozar, A. Shi, L. Yang, S. Shirowzhan, D.J. Edwards, Additive manufacturing applications for industry 4.0: a systematic critical review, *Buildings*. 10 (2020) 231. https://doi.org/10.3390/buildings10120231.

[53] C. Li, H. Lei, Z. Zhang, X. Zhang, H. Zhou, P. Wang, D. Fang, Architecture design of periodic truss-lattice cells for additive manufacturing, *Additive Manufacturing*. 34 (2020) 101172. https://doi.org/10.1016/j.addma.2020.101172.

[54] F.M. Ribeiro, J.N. Pires, A.S. Azar, Implementation of a robot control architecture for additive manufacturing applications, *Industrial Robot*. 46 (2019). https://doi.org/10.1108/IR-11-2018-0226.

[55] M. Khorasani, A.H. Ghasemi, B. Rolfe, I. Gibson, Additive manufacturing a powerful tool for the aerospace industry, *Rapid Prototyping Journal*. 28 (2022) 87–90. https://doi.org/10.1108/RPJ-01-2021-0009.

[56] M. di Nicolantonio, E. Rossi, P. Stella, Generative design for printable mass customization jewelry products, In: *Advances in Additive Manufacturing, Modeling Systems and 3D Prototyping: Proceedings of the AHFE 2019 International Conference on Additive Manufacturing, Modeling Systems and 3D Prototyping*. Springer International Publishing, Washington DC, 143–152, July 24–28, 2020. https://doi.org/10.1007/978-3-030-20216-3_14.

[57] T. Spahiu, A. Manavis, Z. Kazlacheva, H. Almeida, P. Kyratsis, *Industry 4.0 for fashion products - Case studies using 3D technology*, IOP Conference Series: Materials Science and Engineering, IOP Publishing Ltd, 1031(1), 012039, 2021. https://doi.org/10.1088/1757-899X/1031/1/012039.

[58] N. Fatma, A. Haleem, S. Bahl, M. Javaid, Prospects of jewelry designing and production by additive manufacturing, In: *Lecture Notes in Mechanical Engineering: Select Proceedings of ICRAMERD 2020*, 869–879, 2021. https://doi.org/10.1007/978-981-33-4795-3_80.

[59] H.C. Koch, D. Schmelzeisen, T. Gries, 4D textiles textiles-an made by additive overview manufacturing on pre-stressed textiles-an overview, *Actuators*. 10 (2021) 31. https://doi.org/10.3390/act10020031.

[60] A. Le-Bail, B.C. Maniglia, P. Le-Bail, Recent advances and future perspective in additive manufacturing of foods based on 3D printing, *Current Opinion in Food Science*. 35 (2020) 54–64. https://doi.org/10.1016/j.cofs.2020.01.009.

[61] M. Javaid, A. Haleem, Using additive manufacturing applications for design and development of food and agricultural equipments, *International Journal of Materials and Product Technology*. 58 (2019) 225–235. https://doi.org/10.1504/IJMPT.2019.097662.

[62] M. Javaid, A. Haleem, R.P. Singh, R. Suman, S. Rab, Role of additive manufacturing applications towards environmental sustainability, *Advanced Industrial and Engineering Polymer Research*. 4 (2021) 312–322. https://doi.org/10.1016/j.aiepr.2021.07.005.

Index

For Product Safety Concerns and Information please contact our EU
representative GPSR@taylorandfrancis.com
Taylor & Francis Verlag GmbH, Kaufingerstraße 24, 80331 München, Germany

www.ingramcontent.com/pod-product-compliance
Lightning Source LLC
Chambersburg PA
CBHW060351220326
41598CB00023B/2881

9 7 8 1 0 3 2 5 2 3 9 9 6